To Rick – a

years! Enjoy

Rick

I Know a Little About a LOT OF THINGS

A Chronical of my Life in Construction

RICK TOUGH

One Printers Way
Altona, MB R0G 0B0
Canada

ww.friesenpress.com

Edition — 2022

6-7 (Hardcover)
-0 (Paperback)
(eBook)

OBIOGRAPHY, PERSONAL MEMOIRS

le by The Ingram Book Company

The photograph on the front cover was taken by me from one of the tee boxes at the Shadow Mountain Golf Course near Cranbrook, BC during a Mother's Day Golf Tournament in June of 2016. Not only is this a fantastic vista of the beautifully manicured fairway with a background of the majestic Rocky Mountains, it beautifully showcases the McPhee Bridge crossing the St. Mary's River. This is a very special view to me as this was my first project as a superintendent in 1980 while working for Manning Construction, at the age of 25-26.

This book is dedicated to
Derek, Taralyn and Chaylene.

Table of Contents

The Dunning–Kruger effect is a cognitive bias stating that people with low ability at a task overestimate their own ability and that people with high ability at a task underestimate their own ability. — Wikipedia

The more you know, the more you know you don't know. — Aristotle

The less you think you know, the greater your ability to learn and grow. — Hal Elrod

As we know, there are known knowns; there are things we know we know. We also know there are known unknowns; that is to say we know there are some things we do not know. But there are also unknown unknowns—the ones we don't know we don't know. — Donald Rumsfeld

It ain't what you don't know that gets you into trouble, it's what you know for sure that just ain't so. — Mark Twain

I know a little about a lot of things. — Earl Fatha Hines

CHAPTER 1 —
My Family Tree

MY FATHER, WILLIAM ALEXANDER (BILL) Tough, was born in Vancouver on November 6, 1924, to Alexander and Alberta Tough. My paternal grandfather, Alexander Beaton, was born on August 16, 1898, at 1:00 AM, at 60 Baker Street, Saint Machar Parish, Aberdeen, Scotland. My paternal great-grandfather, who was also named Alexander Beaton, was born in Aberdeen in 1870 and worked as a journeyman mason. My paternal great-grandmother was Bathia Bannerman, a domestic servant who was born September 1, 1873. Alexander Beaton Sr. did not stay in his relationship with Bathia and rather quickly left her and their child to marry another woman. Bathia continued working and she and baby, Alexander John Beaton Jr. (Alex), lived with her older sister Barbara and her husband, James McRae. Alex was eventually adopted by the man who married Bathia, George Tough, who was born on January 19, 1862. They married in the city of Aberdeen on July 10, 1902. A new beginning took the family from Aberdeen, Scotland to Chilliwack, British Columbia in 1903, where George purchased 43 acres from the estate of John B. Hagen at 10835 Chapman Road in the community of Rosedale, just east of Chilliwack. Four more children were born to Bathia and George on that farm, George Jr., Mary, William and Beth, but mother Bathia died in childbirth with Beth on June 22, 1912. George was left with five children to care for on his own, so he decided to return to Scotland with George Jr., Mary and William. Beth, who was a newborn, was left in the care of the Hamilton family on a neighbouring farm, and in 1913, the Tough family's Rosedale property was sold to William D. Muir. It is assumed that Alex, the oldest child, was also left with the Hamilton's but records hint that he moved to Vancouver at the age of 15 to live independently. George Tough Sr. later returned to Canada with George Jr., William and Mary.

Alex made his start in 1914 as a clerk at Woodward's and lived at different boarding houses: 4531 Prince Albert, 1020 Harwood Street, 420 Davie Street and 89 Pandora. He ran a horse-driven delivery service for Woodward's, Nabob Coffee and worked for Kelly Douglas in the shipping department as well as FF Walker Garbage and EA Morris Delivery until he became a teamster working at Malaspino Fuel Co. Alex met Alberta Lee Houkes (possibly while working as a delivery driver for Woodward's), and they wed on December 12, 1923, in Vancouver. Alex was listed on their certificate of marriage as employed as a cartage foreman while residing at 876 Granville Street, and Alberta was listed as a salesperson residing at 966 Burrard Street. That certificate was witnessed by Gladys Houkes, my great aunt, who also resided at 966 Burrard, and George Tough Jr., my great uncle who resided at 198 W 7th Avenue.

My paternal grandmother, Alberta Lee Houkes was born in Winnipeg, Manitoba to Albert and Amelia Houkes on May 11, 1903. Albert Houkes was born January 4, 1868 in Bradford, England and Amelia was also born in England on May 19, 1875. Albert apprenticed as a stone cutter before immigrating to Australia. He moved to Canada in 1890 and settled in Ontario. Soon after that, he moved to Brandon, Manitoba and married Amelia Charlotte Harland in 1894. After moving back to Winnipeg in 1895, Albert entered into the business of Hooper, Houkes & Company, a marble and granite company. They went on to have a family, starting with Lily in May 1895, Arthur Edward in 1896, Gladys in 1900 and Alberta Lee in 1903. The Houkes family eventually moved out west to Vancouver, settling at 539 Hornby Street. Albert was employed in the marble and granite business with Patterson, Chandler & Stephen. Lily, Gladys and Alberta all started work for Spencer's Department Store on Hastings Street and Arthur became Private Arthur Houkes of the 1st Canadian Mounted Rifles

(Saskatchewan) and lost his life on September 28, 1916, in WWI at the age of 20. He was buried at Vimy Memorial, Pas-de-Calais, France.

Alex and Alberta had two children in Vancouver. My dad, William Alexander, was born November 6, 1924, and my aunt, Shirley Alberta, was born September 19, 1927. The family moved to Penticton, B.C. in 1929 when my dad was five years old, and my granddad Alex bought the Shell Oil bulk distribution plant on Wade Avenue in Penticton after looking after the business for a friend. The distribution plant received deliveries of petroleum products via the Kettle Valley Railway, offloaded the products into storage tanks, then distributed them by small tanker trucks and 45-gallon barrels around the southern Okanagan. Fuel was delivered to gas stations and stove oil was regularly delivered to businesses and residential addresses throughout Penticton as well as Kaleden, Okanagan Falls, Olalla, Keremeos, Oliver and Osoyoos. Grandad ran the bulk oil plant until about 1962, when Shirley's husband, (my uncle) Jim Beasom, took over the plant.

Grandad was a councillor for the City of Penticton and an active member of the local Gyro Club and the Board of Trade for many years. He was also personally responsible for the construction of the famous Penticton sign on Munson Mountain in 1937. Along with others from the Board of Trade, he hauled bags of white silica stone from the Gypo Silica Mine in Oliver on his Shell Oil flat-deck truck and placed the material in the letters of "PENTICTON". He was also instrumental in getting the very first Peach Festival underway in Penticton in 1947 to try and put the city on the tourist map, and that annual festival still operates every summer.

The family lived at 526 Ellis Street (one of the few original homes left in the area today), and Bill and Shirley went to Penticton High School. Apparently, my dad owned a horse in those days and rode it to school, although the actual facts of that story are somewhat vague, as they only lived a few of blocks away from Penticton High. During high school, Dad worked for my grandad at the bulk plant, and one of his jobs after school was to rinse the stove oil out of the 45-gallon barrels and dispose of it onto the ground. To this very day, that lot on the corner of Wade and Powell is used to store new vehicles for an automobile dealership as the land is unsuitable for development due to the enduring hydrocarbon contamination, 80 plus years later!

Once Dad graduated high school in 1942, he took the Kettle Valley Railway to Vancouver with his friend Don Coy and enlisted in the Canadian Navy to join the war effort. He was stationed in Esquimalt on Vancouver Island and patrolled the North Pacific from Victoria to the Aleutian Islands in Alaska on ships called corvettes. A corvette was a small, lightly armed Canadian warship used for anti-submarine warfare in WWII and was about 190' long and only 33' wide[1]. I don't know much about Dad's time in the navy except that he seemed to enjoy the odd kitchen duty and brought home a number of basic recipes from those days, including potato pancakes, eggs fried in bacon grease and steak bread. I also recall a story about one of his shipmates who was washed overboard during a storm, which I believe was the only fatality that he was involved in during the war.

After the war was over, Dad returned to Penticton and went back to work for my grandad Alex at the Shell Oil bulk plant. Some of my dad's friends in Penticton in those days were Al Kenyon, who worked for his father at Kenyon Contracting; Don Coy, who eventually became a butcher in Oliver; and Soapy McIntosh, who later ran a machine shop in Penticton. Dad once told me that after the war, he worked with Soapy McIntosh on some of the new post-war houses in Penticton up in the 3K neighbourhood, which was short for Killarney, Kilwinning and Kensington Streets. Interestingly enough, I owned one of the houses in that neighbourhood as a rental property some 60 years later.

My nana Alberta Tough died on October 20, 2002 at the age of 99. My grandad Alex Tough died on May 8, 1981 at the age of 83.

My mother, Lillian Mary McKenna, was born in Kaslo, B.C. on January 9, 1924, to parents Henry Bernard McKenna and Sarah Katherine (Sadie) MacPherson. My maternal grandfather, Henry, was

---

1 My life and career spanned the period of time where Canada stuck one foot into the metric system while still clinging to age-old imperial measurements. My use of imperial or metric measurements in this book follows whatever the current system was at any given time. My editor strongly urged me to convert everything to metric, but I resisted. I still cannot go into the lumber store and order a 38mm x 89mm instead of a 2" x 4".

born in Queens County, Prince Edward Island on March 1, 1893, and died in Murrayville, B.C. on August 27, 1971. Henry was a bookkeeper for numerous mines and businesses operating in the Kootenay District and the Pacific Northwest US. My maternal grandmother, Sadie, was born in Bonners Ferry, Idaho on June 23, 1901, and died in Surrey, B.C. on March 2, 1998. Sadie's mother, Elizabeth "Lillian" Cassell, was born in Berkshire County, England on June 9, 1873, and her father, John Lauchlin MacPherson, was born in Valleyfield East, Prince Edward Island on November 15, 1874. Family records for Sadie's paternal side go back through Prince Edward Island and Scotland to 1805 and on the maternal side they go back to 1754 in Berkshire, England.

Henry and Sadie had six children: Lillian, Edith, Kathleen (Kaye), Sheila, Theresa (Joan) and Jimmy. Lillian, my mother, was the oldest and was born in Kaslo on January 9, 1924. Edith was born in Nakusp on July 14, 1925; Kaye was born in Creston on July 31, 1927; Sheila was born on November 26, 1929, in Helena, Montana; and Joan was born in Rossland on December 14, 1931. The youngest child, Jimmy, was born in Rossland on March 19, 1933. My mother went to high school in Rossland and graduated in 1942. While living in Rossland, Lillian, Kaye and Sheila became avid skiers at Red Mountain, and the rest of the family no doubt skied as well. Rossland was such a ski town that my mother actually used to babysit Nancy Greene there, who was probably the most famous Canadian woman skier of all time.

After graduating from high school, my mother took nursing training at St Joseph's Hospital in Victoria, then at the Penticton Regional Hospital on Haven Hill Road in Penticton. At one point, in about 1948, Bill Tough was at a local dance in Penticton and met Lillian. They were apparently awestruck and married one year later on January 10, 1949, in Chilliwack, B.C. They moved to Burnaby in 1950. My mother continued her nursing career at St. Paul's Hospital and Vancouver General.

Mom and Dad

CHAPTER 2 —
Ricky's Early Life

I WAS BORN ON APRIL 6, 1954, at Vancouver General Hospital while our family lived on East Hastings Street in North Burnaby. One of my first memories, and coincidently, my very first brush with the law, was when I was about two or three years old and we lived on Sardis Street in Burnaby. One day, my older brother Bill and I decided we were going to walk by ourselves to Simpson Sears up on Kingsway to look at all the Christmas displays in the toy department. I have no idea what my mother was doing at the time, but I'm sure she thought we were just playing cowboys and Indians in the neighbourhood, like usual.

Now, I can only assume we had been to this toy department with our parents at some previous point in time, but regardless, we made it there, and I remember looking at all the toys, in particular, the train sets running around and around. However, something went awry and we somehow got split up, and I ended up attempting to walk home by myself. In the end, a police car brought me home and everything turned out fine. I have no recollection about what happened to Bill, and I assume that he made it home without the police's assistance. Based on Google Maps, I estimate that the distance from the toy department to my home was approximately 1 km. It was quite the walk for a two-year-old with a four-year-old as his escort, especially since we had to cross over a busy street like Kingsway!

Ricky and Billy with Santa

In about 1957, our family moved from Burnaby out to the boondocks in Surrey to a two-bedroom rancher on a decent-sized lot on 89A Avenue, between 132nd and 134th Streets. People at my dad's work at Shellburn Refinery in Burnaby thought he had gone nuts going that far away across the Fraser River. Of course, this was before the Port Mann Bridge and the 401 Freeway were built, so the only crossing was the Pattullo Bridge. Even the Massey Tunnel was not completed at that time.

I recall that my dad built all of the bedroom furniture for Bill and I in the carport from fir plywood and painted it all the same light bluey-green colour. My dad also seemed to like building rock walls in those days. He used a borrowed concrete mixer for the mortar and built walls down each side of the driveway and for the raised garden areas in the backyard. Another memory I have of that house was building forts in the backyard from building materials salvaged from neighbouring homes being built. Those were great places to play cowboys and Indians or hide-and-seek. In those days, there was an open hayfield with dairy cows from 89 Avenue right over to 88th Avenue, so that would also provide a lot of open space to play during the day but you had to watch your step! There was also a bush down by 134th where we and the neighbour kids spent a lot of time running through the forest, climbing trees, etc. The kids who lived next door to us were Ricky and Gary Goodrich. Ricky was my age and Gary was Bill's age, so we hung out quite a bit. I was a bit taller than Ricky Goodrich, so he was known as Little Ricky and I was Big Ricky in that neighbourhood. Another kid a few homes down the street who was about our age was George Hansford. In later years, we all played Little League together, even though we had moved to North Delta by that time. Interestingly, in about 1999, I was boating in the Gulf Islands and I pulled into a dock on Mayne Island and started talking to a guy who was already tied up there, barbecuing, and it turned out to be George Hansford of all people.

Road Trip

In the summer of 1959, our family was travelling to Penticton in our blue 1950 Chev two-door sedan for our annual summer holidays when the engine started knocking badly while climbing the hill at Similkameen Falls on the Hope-Princeton Highway. Since there was a Shell gas station located at the top of that hill in those days, my dad quickly pulled in and drove the car right to the edge of the bank, which dropped down to the falls several hundred feet below, and shut off the noisy engine. Our family had grown that April with the birth of my sister Kathy, and on that trip, she was only three or four months old, Michael was two and a half years old, I was five and Bill was seven years old—we were quite a large family to be riding in a 1950 Chev two-door sedan, but that was the way we rolled in 1959! Aside from the tight squeeze and the diaper bucket, the fact that both my parents smoked like chimneys added to the ultimate pleasure of road travel in those days, never mind the fact that I regularly suffered from motion sickness. Well, it turned out that the engine was totally shot and we no longer owned a car, so we left it at the edge of the bank. Later on, we often wondered if the proprietor had just pushed it off the bank into the river as the car was basically worthless at that point. (I would love to have that car now for a complete restoration.)

Me in Front of the 1950 Chev

Now we had the logistical problem of being in the middle of nowhere—should we somehow return to our home in Surrey or somehow carry on to Penticton? Well, after some no doubt tense discussions between Mom and Dad, it was decided that we would all hitchhike a ride to Princeton, which was about 30 miles ahead, and off we went. We all piled into one lucky guy's car, and I remember getting "Jimmy Legs" as we drove there because it was impossible to move my restless legs the whole trip. We were dropped off at the Riverside Motel, an old-log-cabin-style auto court located on the Tulameen River that is still there to this day, located right near Main Street and the old timber trestle. The plan was for us to stay at the motel for a few days while Dad caught a Greyhound bus back to Surrey where he could go see the bank manager at the Bank of Nova Scotia in Whalley and get a loan for a new car, then find a suitable used car and return to pick us up. I wonder if my mother wondered if he was ever coming back. Well, eventually he returned and loaded us all up into the green 1956 Plymouth four-door wagon, and we continued on our vacation to Penticton. We all loved visiting Penticton each summer, where we would visit with Grandad, Nana, Uncle Jim, Aunt Shirley and our cousins Patty and Bonnie. We loved spending time roasting our skinny white bodies in the hot sun at the beach, swimming all day long, and having bonfires and cookouts in the evenings.

Okanagan Lake, Penticton

Left to Right - Dan MacLanders, Kenny McPherson, Susan McPherson, Michael Tough, Rick Tough, Bill Tough, Doug MacLanders. Front - Scott MacLanders

One other great part of visiting Penticton in the summers was staying at a rustic old cabin right on Okanagan Lake at Trout Creek point. The cabin was owned by the Hill family who were next-door neighbours to Uncle Jim and Auntie Shirley on the West Bench. It had a woodstove, a handpump for water and an outhouse out behind the cabin. As it had no electricity, we used kerosene lamps inside the cabin at night. The best part of our time at the lake was swimming from the raft.

Once we were a little older, Bill and I slept in Mr. Schirmer's old converted bus out in the woods behind the cabin. He was a buddy of Mr. Hill, and he spent part of his summers on the property. Staying in the bus was always interesting because it was easy for me and Bill to roam around in the middle of the night. I know the Hills didn't charge Dad for our annual use of the cabin, but we always left the place in far better condition than it had been in when we'd arrived, raking the beach before leaving and cleaning the place up. One year, we even re-stained the entire exterior of the cabin.

Cabin on Trout Creek Point

Bill and I both started elementary school at David Brankin, which was located on 88th Avenue and 132nd Street. According to Google Maps, this was a daily walk of 1.4 km each way, but in fact, only one direction was uphill! We either walked or rode our bikes, but we definitely did not get a ride. One of the interesting things about that daily walk was passing by the Pedevich Farm along the way on 88th Avenue. It was an old, dilapidated farm house that looked like it was going to fall down (this was in 1960, and the place is still there in 2020 according to Google Maps), and they had all kinds of junk and wrecked cars spread out around their farm, along with a number of cows. I was only at David Brankin a couple of months before we moved to North Delta. I was at the principal's office (Mr. Heinz) only once during that time, for accidentally running through the girls' gymnasium while being chased in a game of tag. I was required to stand at attention in the hallway outside his office for some period of time while awaiting my fate for this transgression. (Fairly clean record, I would say!)

In those early days, as it likely was for most families, visiting the relatives was very common, and we often went for Sunday drives to see our great uncle George (Alex's half-brother) and Aunty Olive Tough on Dent Street in Burnaby, along with cousins George and Rob, or over to Newton to see Uncle Kent, Aunty Edith MacLanders (my mother's sister) and cousins Doug, Dan and Scot. Grandma McKenna moved to a big old house on Victoria Avenue in White Rock around 1960, so she was very close to visit as well. We spent lots of time at the beach there in the summer having fish and chips in newspaper and burning our white bodies until they were beet red, as we apparently didn't know too much about sunblock or skin cancer in those days. Uncle Norton and Aunty Sheila Young (my mother's sister) and my cousins Noreen and Brent, lived only a few blocks away in North Delta, so they were very close by as well. I remember going on a family ski trip once with Uncle Norton, Aunty Sheila, Aunty Kaye, Noreen and Brent, but we didn't get the hang of skiing too well and we got soaked wearing jeans and real nice knitted wool mitts! Other family trips included riding bikes through Stanley Park with Uncle Norton and Aunty Sheila leading the way (that was a lot of fun) and road trips to downtown Vancouver to visit Great Uncle AE and Gladys (my maternal grandmother's sister) who lived in an apartment in the West End. Sometimes we would visit my Great Aunt Lily (my maternal grandmother's other sister) who lived at 1210 Jervis Street in Vancouver. On a few occasions, we made the long trek out to Chilliwack to visit Great Uncle Russ and Aunty Beth (my paternal grandfather's half-sister). Most of these people had bowls of mints or candies for guests, but we had to mind our manners. (Speaking of manners, quite possibly due to our British heritage, both of our parents constantly drilled us about the proper use of manners in every aspect of life, and it was as if this was taught to us so we could prosper in life and not be considered lower class. We were often asked, “What if the Queen were here?” as if we were being trained in case the monarchy might drop over for dinner at any given time. One of my favourite aunts was Aunt Kaye, my mother’s sister who moved to Reno, Nevada in 1964. Whenever she visited, it was always fun. We stayed close to all of these people over our entire childhood, and they were always a big part of our extended family.

Perhaps my Favourite Family Photo Taken in Pritchard
Left to Right – Bill Tough, Doug MacLanders, Aunty Kaye Oppio,
Uncle Kent MacLanders, Grandma Sadie McKenna, Ricky

In November of 1960, my parents bought a brand-new three-bedroom, two-storey house on 92nd Avenue for $18,000. They were able to get a mortgage for 25 years at a fixed rate of 6%. We had a full basement and a lot more room than in the last house. It wasn’t very long before rock walls started appearing and the driveway was paved with concrete from a borrowed concrete mixer and hand-shovelled gravel and bagged cement. A large sundeck was built onto the back of the house and a smaller sundeck was added onto the front of the house. Two bedrooms were eventually built in the basement for Bill and me, along with a rudimentary bathroom and even a rec room. The homemade furniture survived the move from Surrey and more handmade, built-in bluey-green shelves and a desk were added to each of the bedrooms. Keep in mind that the tools my dad used to construct these things were very limited and mainly included a crude hand miter box, a handsaw and a hammer. Not all angles were perfectly fitted, as one might expect if using a more modern miter saw, but I give Dad an "A" for effort for making the most of what was available to him on a tight budget.

Once we moved to 92nd Avenue, we enrolled in Annieville Elementary—I in Grade 1 and Bill in Grade 3. We quickly met many new friends in the immediate neighbourhood, including Brad and Steve Taylor, Ken Manning, Richard Knight, Ken Hicks, Mathew Heinz, Wayne Lock, Doug and Richard Coulson and many others, as the relatively new neighbourhood was filled with kids our age. Riding bikes, playing and collecting marbles and baseball cards were the games of that early era. Playing in the bush across from our house also became a major pastime for us. In those days, it seemed that my parents were very busy with Dad working shift work at the refinery and Mom doing all the cooking, cleaning and laundry and raising my younger brother, Michael, who was three years old, as well as Kathy, who was one year old. That was like a free pass for Bill and me because we could leave in the morning on our days off from school and just roam all day long as long as we were home for dinner on time. Within months of moving to our new community, we would ride our bikes down to the Fraser River and play in the sand dunes or go up to Kennedy Park to the swimming pool or anywhere in between for that matter. One hobby that we got into pretty quick once we moved to Delta was building go-carts and racing them down 93rd Avenue, starting at 116th Street, which was the steepest hill close by. Kids would just show up and start racing down the hill in their ramshackle go-carts, sometimes going head-to-head against each other in races. We steered using a rope that was fixed to each end of the front axle and looped through the driver's area. Brakes, if you had any at all other than your running shoes, were often composed of a stick that was bolted to the frame and could be pulled back by the driver so the bottom of the stick would drag on the asphalt below. My dad got quite involved one particular year and built us a hot-rod-style go-cart by using thin-steamed Masonite sheets, which allowed him to roll the material into a hood over the front and a smaller one over the back end, like a trunk. That was a pretty fancy go-cart, and it was all painted bright yellow.

A bit later in our early childhood, we started playing "scrub league" baseball down at one of the fields at Annieville School. We called it the Punk League. It was played with any number of players, as long as we had a baseball bat, a baseball and hopefully nine baseball mitts. The first thirteen players would take the regular positions on the field on a "first come, first serve" basis from catcher, pitcher, first base and so on until right field was filled, including four hitters. Any additional players would have to wait for an opening. Then the game would begin, and the pitcher would pitch to the first batter. If the batter got a hit and got on base, the next batter would be up. If a batter was put out with either three strikes or by a fielder, that batter went to the back of the waiting line and all positions would switch forward one position. For example, the first guy in the waiting line would become right fielder, the right fielder would go to centre field, and the pitcher would become the catcher and so on. The game had no innings, no time constraints, no uniforms and no umpires, as long as reasonable interpretation of the rules of Annieville Punk League baseball were followed. If there was ever a dispute or a tie to be resolved, the matter would be rectified by a bat toss. This was a process of selection that involved tossing the bat to another player, then going hand over hand up the bat, and whoever had the little bit of the butt of the bat to grab was the winner.

The games in Punk League could go on for hours and hours and you could keep on hitting if you were able to avoid striking out or being put out. Players from those days came from all over the Royal Heights subdivision and the wider Annieville area and included names from the past like Joe Pulaski, Bob Akrigg, Bob Tonsaker, Howard Nelson, Lawrence Betts, Steve Taylor, Ken Manning, Andy Millar, Gary Robilliard, Stu Graham, Sonny Rattray, Mike McNeill, Ron McNeill, Ralph Savage, Dale Ramsay, Brian Lindsay, Terry Lindsay, Gord Williams and Bill Tough, to name a few. One sure way to end one of these seemingly endless competitions was when the guy that brought the only bat had to go home. Sometimes, it might be deemed necessary for some of the senior players to offer enticement to that individual to stay longer, like jumping the queue and getting to bat earlier.

Other activities I remember in those days include fishing off the government dock at Sunbury or gathering eulachons on the river's edge. Sometimes we would catch a sturgeon that you couldn't eat, but they looked so weirdly prehistoric. Other great pastimes that were better left unknown by your parents were playing on the log booms on the Fraser River, sitting on the girders of some of the railway trestles down by the river as a train went over it or visiting some of the freighters tied up at the Fraser Surrey docks where you might trade coins with a friendly Russian mariner!

Little League was another great pastime for Bill and me in our early days of North Delta. Early in the spring, there would be an announcement (probably in the local paper) that registration would take place up at Kennedy Heights Shopping Center and tryouts would be held at Kennedy Park. As I was always tagging along with Bill, who was two years older, we were always put on the same team. As far as I know today, we only played for the Cardinals and we stayed on that team for the entire Little League term. Ed McKenzie was the coach and a great guy. He had good knowledge of baseball and was very easy to like and get along with. He took the game seriously and worked hard at teaching all of us the skills of baseball as well as the work ethic required to improve as players and to win as a team. His two sons, Jim and Dave, were also on the team and I recall they played second base and shortstop, respectively. Other players I remember were on the Cardinals were Tim Murphy and John Schreiner. Tim switched up with me as a catcher and later attended BCIT in the same class as me in the mid-1970's.

I remember many of our opponents quite clearly to this day. Dave Coutu, who played for the Pirates, ran me down in a major collision at home plate as I waited for a throw during one game, and I later learned more about his extreme competitiveness playing basketball with him at North Delta High School. That collision might have been my first serious concussion, but there were many more to come. Other players on the Pirates who I remember were Darryl and Robbie Anderson and Rick Cowie. One excellent player we played against was Bob Tonsaker, who pitched for the Orioles. Besides being a great all-around player, he had a terrific fastball with great accuracy. Another notable pitcher we faced was Gord Williams, who also played for the Pirates, and I remember he had a very high windup where the ball came from way up high with speed. He beaned me in the back once when I was not able to get out of the way quick enough. Dale Ramsay was another great pitcher from the Indians who had a wicked curveball for his age. Several other Indians included Gary Robilliard, Stu Graham, Don Clipperton and Dan Neil. What a fantastic group of athletes those guys were! I went on to play Pony League at Sunbury Park for a year or two, but, while I was probably a decent catcher, I did not really excel at batting, and I eventually moved on to other sports. (Remember, in baseball, getting on base 30% of the time is a pretty good batting average.)

I am really glad I played baseball because it is a great game with so much history, and I really enjoyed playing the game, maybe even more so in Punk League than Little League, but in the end, it apparently wasn't really my life sport. Many years later, I read a book called *Dynasty – The New York Yankees 1949 – 1964* written by Peter Gollenbock, and it took me back to that era a long time ago when we were very young, carefree and playing baseball and all the familiar names from major league baseball came streaming back in my memory, many of them from the Yankees organization. It was such a great read and so revitalizing that I bought another copy of the book and on a boating trip in the summer of 2003, I hand delivered it to my old coach, Ed McKenzie, who was then retired and living on Mayne Island. I thought he would enjoy the way the book took you back to a completely different era, mainly because I know how much he loved baseball. Along with the book, I gave him a heartfelt letter thanking him for all the effort he put into coaching us and apologizing for not putting more dedication and effort into the game at that time. Ed also keenly followed our high school basketball team, and he saw the more positive results we achieved through a much greater dedication to that team than I had provided him in Little League. It was a very nice meeting with Ed and his wife, Beverley, but that was the last time I saw him. Unfortunately, he died in 2015 at the age of 81.

CHAPTER 3 —
Early Entrepreneurial Adventures

MY FIRST ENTREPRENEURIAL VENTURE WAS with my brother Bill from about 1957 to 1958, when I was three or four years old. Our parents liked to grow tomatoes in the backyard on 89A Avenue in Surrey, and I guess we had a particularly big crop that year so we loaded a pile up in the wagon and towed it through the neighbourhood selling tomatoes from door to door. Although many of them were a bit green, our sales pitch was to tell the customers to place them on the window sill and they would ripen up just nice. I am not sure what the selling price was, but I am sure that was a nice little bump to the piggy bank.

In later years, once we moved to North Delta, and prior to getting a real job, I did many things to earn money, including shovelling snow off driveways and roofs. Don Clipperton and I actually had a sales pitch about how this necessary roof service would prevent leakage through the roof if we cleared at least the bottom 4' of snow around the gutters. In later years, I delivered the Columbian Newspaper, the Star Weekly and the Vancouver Sun, which were all afternoon deliveries. Thank God I never had to deliver The Province newspaper because that was a morning delivery, and I would have had to get up at something like 4:00 AM to be finished before going to school! Delivering papers at such a young age to many various houses and meeting the customers when I collected the money taught me a lot about the diversity of people, at least in our extended neighbourhood. Some people had beautiful lawns, gardens and freshly painted picket fences, while others just let the weeds grow. Some people were nicely groomed and pleasant while others came to the front door in dirty wife beater shirts. I often equated those traits and behaviours to pride. Some had it, others didn't. I came to admire those people with pride.

A couple of summers, we caught a dirty old school bus from Kennedy Heights Shopping Center over to Driediger's Farms in Langley to pick strawberries to try and make a few extra dollars. This was the worst job of my life as I hated the smell of the sweet, sticky strawberries and I cannot stomach to eat one single strawberry to this day. One summer, I would guess in 1966, Bill and I got a job picking cherries at my Uncle Jim's and Auntie Shirley's orchard in Penticton. I enjoyed eating the odd cherry as I picked, but by the end of the picking season, I was pretty tired of cherries too. This was a pretty good job, especially because we stayed with my aunt and uncle and Patty and Bonnie, our older cousins. I think it was that summer that Bill and I saddled up their horse, Danny, and rode him up to a big field adjacent to the elementary school on the West Bench. Once we were there, we raced Danny from one end of the field to the other to see how fast we could get him galloping, but he wasn't too good at turning around or taking too many orders from us inexperienced riders, so we sometimes had to literally get off the horse, turn him around by hand, then get back on and race him down the other way. Well, this went on for a few sprints for each of us until the horse decided he was heading back to the orchard while I was on him, racing full speed down the rough macadam road, charging into the orchard and using the cherry tree branches to knock me off in the process. I broke my arm in the fall and ended up in a cast for the rest of the summer. We never rode that stupid horse again, but that wasn't the end of my cowboying days!

In the summer of 1967, when I was 13, I was hired by our neighbour on 92nd Avenue in North Delta, Kenley Adams, to cut his lawn all summer while they took a trip across Canada to celebrate the Centennial in their homemade camper on their Rambler! They had a bit of an acreage mostly covered in lawn and trees, so this was a fair bit of work. Kenley, who had two daughters and no sons, was a particular sort of guy, being an industrial arts teacher, so I had to go through a significant one-on-one training program before they left, although I think Kenley enjoyed having me around to train. The training program included checking the oil before starting and learning the proper methods of fuel refilling, including filtering the gas and topping up the tanks before putting the mowers away. The maintenance program also included training on spark plug cleaning and gap setting, blade removal and sharpening with a hand file (counting the strokes per side) and removal of grass buildup from the lawnmowers. As I recall, there were two separate mowers, a gas-powered push mower and one with automatic drive (that was a first in that era I am sure). The other component of training was on the methodology of lawn mowing. Kenley instructed me to start mowing around the bigger trees and flower beds, mowing in a circular direction around each one, shooting the cuttings outward, then expanding in concentric circles until there was a significant area and eventually the circular areas started to overlap. In the end, I would mow around the perimeter of the property until those straight lines overlapped the circular shapes throughout the yard until all areas were completed. Then the raking would begin. My memory is clear to this day on the training regimen, although I am not crystal clear on the amount of compensation I received for that summer's work. I believe it was $10.00 for the entire summer!

It might have been that same year that my dad was on strike at the Shellburn Refinery in Burnaby, so he had no money coming in, yet somehow, he a got a contract to paint the exterior of the old hardware store called Bill's Hardware on Scott Road near 92nd Avenue. Well, naturally, Bill and I were on the painting crew. It was an old wood building, so it took a lot of hand scraping before the painting could even begin.

A couple of notable family vacations in this era, other than regularly heading to the Okanagan or the Kootenays to see our relatives, included a trip to Barkerville one year and another year, driving all the way to Tijuana, Mexico! The Barkerville trip was in the 1956 Plymouth wagon, which towed a homebuilt utility trailer with all of the camping gear in it. That was an unbelievably long drive up the Fraser Canyon and then on up to Prince George, with some of the road still in gravel. This was the longest trip of my life as I thought the driving would never end. In addition to the crowded car and the dust was the constant cigarette smoke coming from the front seat, which was considered "normal" in those days. The trick when you piled into the back seat was that you selected a seat out of reach of Dad and behind Mom so she couldn't get much of a swing at you when you were whining.

Shuswap Lake Camping with the 56 Plymouth

The trip to California and Tijuana in 1967 was epic. Even though it was a very long trip (about 1500 miles each way), we survived the long days in the car because it was so cool to be going to California and the scenery was ever-changing all the way down. For that trip, we were in the 1965 Galaxy 500 station wagon, which was a lot bigger than the Plymouth wagon. The trip was over the Easter holidays, so it took a frantic pace to travel all that way and back in 10 days, especially because we took the ocean route all the way from the Oregon Coast to San Francisco. We spent a day or so in San Francisco and in Los Angeles and Hollywood, which was fantastic for a 13-year-old kid. The day excursion into Tijuana was in a completely packed taxi, and what we saw there was a huge eye-opener for all of us, including the dirty little kids selling Chicklets and the upholstery shops everywhere. The various memories of that trip will never fade in my mind, and I continue to have a special affinity for palm trees ever since that trip.

The summer of 1967 was also the year that I was very fond of Jacquie Nelson, a local girl who lived in the neighbourhood. She was one grade ahead of me and obviously much more worldly, but we had a lot of fun together that summer riding our bikes down to the sand dunes, and on at least one occasion we snuck out together in the middle of the night and laid on our backs down at the Annieville Park tennis courts and watched the falling stars during the Perseids Meteor Shower. However, that relationship did not last as she was heading off to Grade 8 at North Delta and it wouldn't be too cool to be going out with some kid from elementary school! Notwithstanding my broken heart (my dad had warned me what would happen), Jacquie and I remain great friends to this very day.

In the summer of 1968, both my brother Bill and I were hired by Kenley Adams to assist him in building his brand-new house. We were paid a princely sum of $1.00 per hour for this manly work; however, prior to the start of construction, the first task at hand was to split some large stumps that remained from the lot clearing exercise. This may well have been Kenley's way of providing physical conditioning for Bill and I as he was a bit of a fitness buff and perhaps, he thought we needed a little strenuous exercise before we began working on the house.

Notwithstanding the fact that we had real money-paying jobs, we were still required to perform certain tasks at home. These included cutting the lawns, which included doing the edges with hand shears as well as raking up afterwards and putting all the grass into the compost pile. Every now and then we had to use the hand shears to cut all the weeds and long grass in the ditch that ran the length of our lot along the road. That was not a fun job. Talk about instant tendonitis! What a marvelous invention weedeaters were, but they were invented far too late for 92nd Avenue. My parents were quite proud of their lawns, rockeries and flowers, and our enthusiasm for quality work was not optional. Other household chores that Bill and I often shared in that era included washing the dishes, ironing and folding laundry and even darning holes in socks and sewing on buttons. My mother, like many mothers of that era, was a great cook and she made food for a family of six on a strict budget, shopping for food once every two weeks on payday. That took some meal planning! We learned a certain amount about cooking from both Mom and Dad, but Mom was the type of cook that didn't confine herself strictly to recipes and she would say, "Just a bit of this" and a "Couple of shakes of that", so many years later, trying to recreate her dishes remains a lengthy testing exercise. Over the years,

I have mastered a few favourites, like baked macaroni and cheese, hamburgers with mushroom sauce, meatloaf, mock duck, salmon loaf, scalloped potatoes, etc. I remember that desserts, which were quite common in those days, were sometimes limited to canned fruit like peaches and cherries (some of which she canned herself), but the most despicable was Royal City Canned Plums. Apparently, they were canned whole with the skins on and even the mere thought of that rough hide scraping down my throat would make me gag before I could be excused from the table. Again, leaving any food on your plate was not an option available to us, so we somehow finished those plates, but the process probably caused a bit of emotional damage along the way.

The actual house construction for Kenley Adams began with the excavation of the foundation, which was generally performed by a rubber-tired hoe, but any of the tight spots in the hard-packed, cobbly, glacial till common to that area were hand dug by Bill and me using picks, shovels and wheelbarrows. Once the foundation excavation was complete, we assisted in the construction of footing formwork, which consisted of 2" x 10" lumber and 1" x 4" strapping. The most difficult part of the footings was the pounding of 2" x 4" stakes into the hard, rocky till to secure the forms in place. This was where we learned the finer points in the use of a "bull prick". A bull prick is an iron rod about 18" long and 1-1/2" in diameter with one roughly sharpened end. You use a sledge hammer on the other end to drive the instrument into the hard till, eventually breaking the way for the stake to be installed. Fun times indeed!

The foundation walls were more like real construction because we got to do some sawing wood, hammering and nailing; however, Kenley's choice of materials for this was a low-grade shiplap, which was riddled with knotholes and cracks, so strength and durability was a real concern. I remember a lot of this lumber was severely damaged in the unloading and handling process at its final destination, so who knew how it would stand up when pouring the concrete? The other aspect of these wall forms was the use of metal strap ties. This crude formwork method involved making a small vertical sawcut in the shiplap at specified locations adjacent to a stud, say at 24" centres, and then inserting the strap and nailing it to the stud. This strap was installed to resist the lateral pressure of the fresh concrete. In fact, if my memory serves me correctly, the strap was supposed to wrap around two sides of the stud so the tension in the strap was not in direct shear to the nail fixing the strap to the stud. But in this case, I believe the studs were 2" x 6" and the straps were designed for 2" x 4", so the nailing was on the side of the stud, drastically reducing the capacity of the forms. It was fun to build all the forms but eventually the day came when the concrete had to be poured into the forms. That was a bad day, and Kenley was sweating big time as the concrete was slowly chuted into the wall forms. Nails were popping and strap ties pulling out, rotating the studs, and the forms were expanding with bulges all over the place! With all due respect to Kenley, who has long since departed this world, this was unfortunately a primary lesson on how not to form walls. In any event, we were able to maintain the forms well enough and most of the bulges were buried and not known to anyone until now. Once the concrete cured, we stripped all of the forms off, which was fun, but the amount of form material we could salvage was close to zero due to the terrible condition of the material. In those days, the foundation form material was generally used for wall framing but very little of this material was reusable.

Once the foundations were stripped, one of our first tasks was to construct the perimeter drainage around the entire house foundation, which just happened to be a sprawling rancher, so this was a big job. A constant grade of 2% had to be strictly maintained by using a hand-selected, straight 2" x 4" with a hand level attached to it. Again, this was tough work due to the very hard, rocky granular till, in addition to the spilled concrete everywhere, so this operation involved the bull prick and lots of pick-axing and shovelling. Finally, once the perimeter drainage grading was completed down to the low point of the house nearest to the street, a trench was required to connect that drainage course to the ditch running along the street; however, it just so happened that the pile of soil material from the house excavation was located right in that path! This was another ridiculous situation where a small rubber-tired hoe should have been brought in to dig that all up in a short matter of an hour or two, but when workers were getting paid $1.00 per hour, the option of utilizing cheap labour was too enticing for Kenley to opt for the excavator. The hand excavation through the excavation pile was a massive undertaking because the pile was likely about 10' high, and it literally took several days to complete. My later surveying and engineering studies would have assisted us in this excavation by laying out slope stakes that would define the safe working slopes of the completed excavation. Without those skills in 1968, we would just try and dig the steepest walls possible until they eventually caved in, thereby requiring re-excavation of the entire trench. Digging and shovelling and wheelbarrowing dirt for eight hours a day for days on end is exhausting work for a skinny 14-year-old kid, but my wallet was expanding steadily by eight bucks a day. (My mother used to say that if we didn't do well in school, we would end up being ditch diggers, so this was an experience that I definitely did not want to do for a lifetime profession.)

On one particular occasion, I remember my dad walking over to the construction site to see what we were up to, and Kenley gave Bill and I high praise. He said to my dad, "I don't know what you did, but you sure taught these kids to be good workers!" Praise wasn't thrown around too loosely in our house, so this certainly made us feel pretty good. One thing that my dad taught us well was "if it is worth doing, it is worth doing right" and that lesson remains thoroughly engrained in my brain to this day.

The only relief from our hard labour was stopping for 10-minute coffee breaks in the mid-morning and mid-afternoon and going home directly across the street for a 30-minute lunch, where we would make a broiled cheese and tomato sandwich on Mom's homemade bread, put the headphones on and listen to the Beatles or Rolling Stones at a high volume to rest, unwind and re-energize.

The following season, I worked for Kenley again on the house, but Bill worked in the oil patch around Calgary as a welder's helper with his friend William Whitton. That season's work involved much more carpentry and framing work that was more interesting, less physically demanding and very much more experiential than the previous season's hard labour, and I recall that I was getting $1.25 per hour. Again, Kenley was keen to teach and we would go through planning and mini-training sessions for each and every new operation we undertook. With just two of us working on the house that year, we worked in tandem quite a bit, and I learned a lot about the correct use of the table saw, the radial arm saw, nailing jigs, prefabrication, etc. Generally, it was a fun summer working on that stage of the house. This is when I first started to enjoy the smell of freshly-cut lumber!

However, I also had a second job that summer at Golden Crown Paints that was located at Columbia Street and 8th Avenue in New Westminster, just across from the old train station. I worked on Thursday and Friday nights from 5:00 PM to 9:00 PM and on Saturdays from 9:00 AM to 5:00 PM. My original job there was as a parking lot attendant because the paint store's business relied heavily on their customers' ability to park in their parking lot, but as the store was located adjacent to the Dunsmuir Hotel Beer Parlour, there were often no parking spots left. At the ripe age of 14, I was responsible for enforcing the customers-only parking lot rule, and from time to time I was known to chase down violators into the beer parlour to threaten them with a tow truck. The "Dunnie", as the beer parlour was known, was a smoky, dark place with dozens of round tables with red terry cloth tablecloths and mostly older men sitting and drinking their small glasses of beer. Waiters in white shirts and black vests delivered full trays of beer glasses around the room to the apparently thirsty patrons. Due to my age, I had to dodge the waiters as I chased after my illegal parkers.

Golden Crown Paints with the Parking Lot and the Dunsmuir Hotel

As time went on, aside from managing the parking lot, I started spending more and more time in the paint store (I know, some of you thought I was going to say more and more time in the pub!) where I started to learn about the various types of paint and related products. One job that

I did fairly often was rotate the stock, which involved taking all of the paint cans off the shelves, referencing the date of manufacture on each can, then restocking all the shelves with the oldest product in front or on the top shelves and the newest product in the back or on the bottom shelves. Eventually, I was dealing directly with the customers and giving them advice on whether they needed latex semigloss or oil-based acrylic paint and what brushes and rollers to use. As time went on, I began using the big colour blending machine to make customized colours based on the formulas on the paint chips and mixing the cans up on the shaker. As enjoyable as the job was, they were only paying me $1.25 per hour, which was the same as I made at my heavy labouring job with Kenley Adams. I remember my bosses at Golden Crown Paints were Wally Walker and Mrs. Downie, who I seem to recall was the mother of Blair Downie, who went to North Delta High School as well.

Now, keep in mind that when I started working there, I was only 14 years old and while I knew how to drive, I was not legally permitted to drive, so my only option was to hitchhike each way to and from work. I should mention at this point of the story that I had become proficient in driving a car as each summer we would often drive my dad's car around the property where the old cabin was on Trout Creek Point near Penticton. In addition, I spent time with my mother, teaching her how to drive the old four-speed Volkswagen that she bought so she could go back to work as a nurse. We would drive down to Burns Bog and through all the roads into the bog while she learned proper shifting techniques from me, including double clutching and downshifting, which probably was not taught by the instructors in those days. To further advance my levels of driving proficiency, as several of my friends did at that time, I would simply "borrow" the family vehicle and drive around the country. On one particular evening, I apparently caught the eye of an RCMP officer who was cruising the streets like me that night, and he decided to pull me over for some unknown reason. Well, my parents were pretty quiet that night after they got the call to come and get me and the car at 2:00 AM by Guildford Shopping Center. There was no yelling or incriminations, there was just one, simple statement. That little episode prevented me from getting my driver's licence until my 17th birthday, instead of my 16th birthday. That was enough said.

Well, back to the paint store. Based on Google Maps today, the distance from my house on 92nd Avenue in Delta to the paint store was 9.7 km, which included a walk of 1.1 km to get to Scott Road where I stuck out my thumb, then another 1.6 km walk from the Pattullo Bridge, where my ride usually ended, down to the paint store. On Thursday and Friday nights, I would usually get home after 10:00 PM, depending on how long it took to get a lift. At that age, I was a pretty clean-cut kid so getting a ride did not usually take very long. I met a lot of interesting characters on that commute, to say the least. As I walked the length of Columbia Street six times per week, I got very familiar with New West and often stopped for a snack or a drink at the Royal City Café, went shopping in the big Army and Navy Store or stopped in to look at some shoes with Cuban heels at Copps Shoe Store. For a young guy with a bit of money in his pocket, it was sometimes hard to make it the whole way down Columbia without spending at least some of it! I know my dad was a bit concerned when I told him how much I was enjoying working at the paint store and that I might want to stay on and eventually become store manager!

At some point, I quit working at the paint store, most likely because I was working too many hours on the Adams' house, but rather than leave them a man short, I contacted Kerry Grozier, who was a good friend of mine who lived just behind me on 92ndA Avenue to see if he was interested in applying for the job. He was happy to go down and apply, and they hired him right away. They started him off in the parking lot and eventually broke him into rotating stock and labelling paint cans. Kerry later stole my girlfriend Debbie Guenther, a beautiful blonde girl who lived in Burnaby and who I met at a concert in Vancouver, presumably because he had a car and I was still riding my 10-speed bicycle! (Surprisingly, Kerry and I are still very good friends to this day.)

Over the next couple of years while I was waiting to get my driver's licence, I did a lot of riding on my 10-speed bicycle, and I loved the feeling of the speed and mobility it gave me. The 10-speeds certainly were a huge improvement over the previous bikes I'd had; they were lighter and so much easier to propel. Brad Taylor and I used to make regular rides into Vancouver to visit his uncle Ken Griffiths who lived at 54th and Kerr, then we would head down Kingsway to downtown Vancouver and back home to Delta in one day. On at least one occasion we went right across the Lions Gate Bridge, through North Vancouver then back across the Second Narrows Bridge and back home. The toughest part of those rides was climbing Scott Road Hill in Surrey when we were almost home. The longest ride that we ever went on was all the way to Cultus Lake one summer weekend, and we camped once we got there. I think that was over 50 miles long. Luckily, Brad's mother Joyce came and picked us up on Sunday afternoon so we didn't have to ride the whole way home. I certainly felt that ride in my muscles, even as a 16-year-old.

In the summer of 1971, when I was 17 years old, I was driving a bone white 1961 Comet that likely had the ugliest rear end in American automobile history. Chevy knew how to make fins, and even Buicks did, but Ford had no clue. It was my first car, and I purchased it for $95.00 at Lester Motors on Kingsway in Burnaby. I soon discovered that the car had a nasty leaking rear main seal that dripped oil from the six-cylinder engine off the flywheel housing until it blew onto the muffler and tailpipe, leaving a smoky trail! When I regularly pulled into the Mohawk Gas Station on Scott Road and 92nd Avenue, I merely checked the gas and filled up the oil, with 25-cent quarts of recycled oil. Notwithstanding all of the above, the car was my transportation and actually, other than the leaking rear main seal, which certainly was not uncommon in those days, that baby ran pretty fine.

That summer, I worked for the Municipality of Delta Engineering Department as a surveyor's assistant, or rodman, as they called it in those days. My friend Don Clipperton's girlfriend at that time was Lorette Lloyd, and Lorette's mother, Ruth McLean, worked in Delta's engineering office, so she was able to put in a good word for me for that summer job. Associated Engineering had a contract to perform engineering for the Fraser River Flood Control Program in Delta and it turned out the Municipality of Delta had an agreement with them to provide suitable labour to assist their surveyors. That multi-level government program was essentially to raise the level of all of the Fraser River dykes in the municipality above the 1948 flood level in order to reduce potential flooding risk for the future. Throughout that entire summer, I worked for either Hank Melgard or John Dagenais as we surveyed the entire perimeter of Delta, including every drainage ditch entering the Fraser River to the north, the Gulf of Georgia on the west side and Mud Bay on the south side. The survey consisted of running level loops and closed traverses between all main benchmarks along the perimeter, then producing accurate location surveys and profiles of all existing dykes, ditches and waterways. Completing those tasks required regularly going onto private property and that usually involved me going to the front door and advising the homeowners what we were doing. Some of those rural houses along the river were eye openers. On one particular day, we were working just west of Ladner and I approached the old, dilapidated farmhouse of Chung Chuck, the famous potato grower, and when the door opened, I clearly noted that the inside of the house had dirt floors. Chung Chuck worked his fields with his two large daughters and on occasions when they were watering the fields, he would blow a loud whistle, which would signal his daughters to turn off the water and run out and add an extension to the irrigation hose. Then the whistle would blow again and the water was turned on again.

This job involved a lot of slashing with a machete or Sandvik for the survey lines through brush and blackberry bushes for all of us; but it also involved a lot of actual survey work for me like waving the rod for levelling, pounding survey stakes into the ground and measuring distances with the steel chain. Throughout the summer. I was provided with lots of opportunities to advance my skills by running the level, setting up the transit and booking readings and notes in the survey book. The methods of surveying and booking notes are universal, so this was an excellent introduction to survey technology, which I would later add to.

That fall, as I headed back to Grade 12, I put the Comet up for sale in the newspaper classified ads because it used so much oil and smoked so badly. I quickly sold the car for $110.00, a tidy little profit of $15.00, without spending any money on repairs. My understanding at the time was that to replace the rear main seal would have required pulling out the engine and transmission, which would have ended up costing much more than the vehicle was worth. A gentleman came out from Vancouver who wanted to buy it. He looked it over after dinner one night and the car was sold. Later that night, I received a telephone call from him and he was quite unhappy with the smoke the car produced from the leaking rear main seal. Having just completed Law in Grade 11 the previous spring, I was familiar with the Latin axiom *caveat emptor* that means "let the buyer beware." Similar to the phrase "sold as is", this term means that the buyer assumes the risk that a product may fail to meet expectations or have defects. I told him of this premise of common law and said that he had a responsibility to look the car over more thoroughly and his response was that he would never buy a car at night again!

My next car was a beautiful white 1963 Ford Galaxie 500 four-door sedan that I purchased on October 8, 1971, from a man on Dunbar Street in Vancouver for $320.00. This car was a 352 cu in V8 with automatic transmission and was in showroom condition. It had the full-sized rubber floormats and all of the seats were covered in plastic bubble wrap. Even the trunk compartment was super clean. That car was a dream to drive with tons of smooth power. I actually took that car out one day with Brad Taylor to Roberts Bank causeway, which was a long, straight section of new pavement, and got it up to 100 MPH. I was riding in style that fall, picking friends up for basketball practice and heading over to Boston Pizza in Whalley in luxury.

A bit later that fall, I was driving in the West End of Vancouver with my girlfriend Shawyn on Halloween night, when all of a sudden, a car came racing around the corner right in front of me and ran directly into my front end. The police attended the accident scene and found me at no fault in the accident, but that was the end of the Galaxie. It was declared totalled because it would cost more to repair than it was worth. Interestingly enough, I found out later that Dean Bros. Collision purchased the wreck, repaired it and used it as a convenience car for its customers. What a shame.

On November 30, 1971, I purchased a white 1963 Ford Falcon two-door convertible for $375 from a man in Burnaby. It had a six-cylinder engine with a "three on the tree" transmission. This was another beauty with a bright red interior and a black convertible top. Everything worked on the car, it was a pleasure to drive and I sure was looking forward to a summer with it. However, on about December 20 of that year, I was driving in downtown Vancouver with my buddy Brad Taylor after shopping at a place in Gastown that was having a big Christmas sale, when another driver turned left in front of me in the intersection of Main and Cordova, striking us head on! This collision had pretty serious impact, launching Brad's head through the windshield and driving both of our knees into the thick metal underside of the dashboard, leaving four large indents in the metal and bending the old-fashioned light and wiper switches located there. Luckily, we were not seriously hurt. As this accident happened next to the Vancouver Police station, there were uniformed police walking on the sidewalk who witnessed the collision and again found no fault on my part. However, this was getting tiring, buying two nice cars and totalling both of them in the space of less than two months. My dad had a simple, but rather insightful, suggestion that seemed to make some sense to me: "Maybe you should stay out of Vancouver for a while." Brad and I, having no transportation to get back home to North Delta that night, went out for Chinese food in Chinatown and then caught a bus back home with a few bruises and some aches and pains.

About one month later, on January 21, 1972, I replaced the Falcon with a light blue 1965 Chevy II, two-door hardtop, which I bought privately from another man in Vancouver for $385.00. It was nice, clean car with a 283 cu in V8 and an automatic transmission. I liked that car, but the most memorable thing about it was that I had to remove one of the exhaust manifolds up at Millar's Garage to replace one of the spark plugs. This was rather daunting, considering all of the manifold bolts were badly corroded from the heat. By the way, Millar's Garage was an automotive shop located on Scott Road near 90th Avenue that was owned by brothers Merv and Ken Millar, who were the father and uncle of Andy Millar, a good school friend of mine. The story was that Merv and Ken won an Irish Sweepstakes prize and they used the money to build this garage. It was a great place for us young guys to either get work done on our cars or, in some cases, we could just use a hoist and the facilities to self-perform the mechanical work. Merv and Ken were very friendly, even though they had a business to run and paying customers to deal with and keep satisfied. The lunchroom in the garage was completely covered in Playboy centerfolds and you could help yourself to a cold beer out of the fridge if you put 10 cents in the tin can nailed to the wall beside the fridge. This was a great place to hang out and get some work done on the cars, and it was quite the meeting place for years. One other memorable mechanic who worked there was Pete Kreek.

He was a nice guy, but I think Pete's breakfast started out with a beer and he continued drinking throughout the day in that same fashion. I would bet his mechanic's skills were much better at 8:00 AM than at 4:00 PM.

For some reason that is quite unclear to me now, I purchased a 1964 Triumph Spitfire roadster on February 9, 1972, only 19 days after buying the Chevy II! It might have been that I thought I could make some money selling the Chevy II or perhaps I had a hankering for something a little sporty (my brother Bill had a Datsun sports car around this time, I think) but in any event, I bought the car from a car lot in North Vancouver for $475.00. I really enjoyed that car, and I certainly put it through its paces. It was basically a go-cart with twice the power. This was an easy car to tinker with as the hood hinged from the front end right up and out of the way. It had an inline four-cylinder engine with a four-speed manual transmission and twin SU carburetors. The main problem that I remember with it is that it had excessive corrosion in the floor beneath the driver's seat, so much so that if the carpet was moved slightly, you could see the ground flying by beneath you. British cars in those days were not known for quality builds. The other problem was caused by some apparent corrosion inside the gas tank where rust particles would collect in the inline gas filter and the engine would choke of fuel. Not wishing to undertake the major hassle and expense of replacing the fuel tank, the easy fix was just to pop the hood, pull off the rubber gas line at the filter and blow the gas back into the tank. That entire maintenance operation could be completed in less than one minute and one time that exact procedure had to be undertaken while holding up traffic on the Pattullo Bridge! Nevertheless, I was 18 years old and was having lots of fun in that old sports car. That was my fifth car that I owned in that year. Even stranger than buying the Triumph, at some point in time later on, I actually traded the Triumph for a 1963 Corvair van. The van had some issues to say the least, and I did not own it very long, but it was my sixth car in that year.

Around that time, Shawyn's parents hired me to finish off the basement in their family home in North Delta. That job involved framing the walls and enclosing the heat ducting, Gyproc installation, taping, filling and painting. I even rented some equipment to install a stipple ceiling that was quite popular in that era! Vic Seder reminded me recently that he helped me on that job! Other work that Bill and I did in those days was stripping foundations on new house construction as well as a couple of concrete bunkers that were built in the Fraser Valley to hold spent grain from the breweries. A local school friend, Mike Wolzen, and his family were involved with those bunkers as they did a lot of trucking for the breweries, so that's how we got that work.

According to employment records that for some reason I still have today, I worked almost full time as a survey rodman for Associated Engineering in the spring of 1972, specifically from April 24 to June 13, when all the while I was supposed to be completing my Grade 12. I recall some of my teachers commenting that I could have achieved better marks if my attendance was better, and I guess this explains that pretty easily. I know my frame of my mind at that time was that I had no intention of attending university right away, so what did it matter whether I got a B or a C+ in a particular subject? I was much more interested in gaining experience and working to get into a higher paying job with more responsibility while making some badly needed money. Even though I always had part-time jobs throughout high school, I had experienced the feelings of poverty, being broke and not being able to pay for things or do things that I wanted to do, and I never wanted to feel like that again. I was driven to work and to earn money, but little did I know at that time, I would experience this terrible feeling of poverty once again in the not-too-distant future!

CHAPTER 4— North Delta High School Basketball

WHILE I WAS STILL AT Annieville Elementary School, I enjoyed playing all sports, mainly the unorganized community or neighbourhood sports. We would play touch football, scrub baseball, or whatever other game someone invented for hours on end in the Samaloff/Adam's grass field across from our house on 92nd Avenue. At a young age, I played tennis down at Annieville Park with my mother's catgut racquet from her childhood and really enjoyed the game. It quickly became one of my favourite sports as it was purely mano a mano and you could mix up your shots left, then right, hard, then soft and make your opponent run. I loved playing with Rick Hanson, a classmate of mine, because he was pretty good, but he liked to lob the ball every now and then as a defensive shot, so I perfected the body shot, smashing it as hard as I could directly at the centre of his mass.

I played soccer in the community for a number of years but never really appreciated the game that much. It just seemed to me that a number of people were running around, often in the freezing cold or rain, chasing the ball and trying to get at it. I had good, dedicated coaches like Coach Wilson and Coach Rainey, who tried to teach the concepts, the positions and the plays of the game, but I never really loved soccer. Oddly enough, I played with some very good soccer players like Dan Neil, Pat Rhola, Wayne Trafton, Glen and Ian Hilder, Bill McRae, Vic and Richard Seder, Dan Chapman, and even Tony Chursky, who went on to play for the Seattle Sounders in the professional soccer league, but this "beautiful game" clearly was not my passion. I also dabbled in community football for a year or two, playing right end for coach Bob Gough, who was a great guy and my next-door neighbour on 92nd Avenue at the time. My job was to line up on the line of scrimmage on the far-right side and run up the field, then look back hoping to catch a rare pass from the quarterback, Mike Gammon. Most times, chances were that the pass would be a bit off and while I was reaching for the ball and looking back, I would very likely be smoked by a defenseman who had me lined up and clearly had the advantage because he could face me and not have to look behind himself like I did. The roughness of and collisions involved in the game didn't bother me too much because the padding provided a lot of protection and impact resistance, but again, I did not excel at this sport and the game did not become my passion either.

IOOF SOCCER TEAM ~1962 Coach McKenzie, Coach Wilson
Top Row Left to Right: Keith McKenzie, Ron Binnington, Curtis Miller, Brian Smith, Lyle Baker, Rick Tough, Glen Deros, Rick Cowie, Paul Huesken
Bottom Row Left to Right: Randy Stinson, Jimmy O'Brien, Dave George, Wayne Trafton, Dan Neil, Terry Jones, Jim Peacey

One day, while I was in Grade 7, the basketball coach from North Delta High, Stan Stewardson, came to Annieville and spoke to the students about wanting to play basketball once we got to high school. I was somehow drawn to that sport, partially because we used to like to goof around playing hoops around school or at parks, but more so I think, because of the status that came with playing basketball for your high school in front of a crowd on Friday nights. I knew guys like Bob Akrigg, Bill Tonsaker, Ron McNeill, Bill Edwards, Glen Foreman and Ken Manning, who were older and played basketball for the senior boy's team, and that appealed to me greatly. Of course, when I got into Grade 8 at North Delta, I immediately tried out for the Grade 8 basketball team. Alf Clark was the coach, and I also had him for homeroom teacher. Alf was a young, handsome, well-tanned guy who was obviously a jock. It turned out he spent the summers in university as an outdoor swimming pool life guard, hence the tan. As it was Alf's first year teaching high school, it was a damn good thing we had Vicki Wied in our math class that year to get Alf through it (Alf would certainly agree, as he told the story many times himself over the years). Alf was an enthusiastic coach and loved to teach us the sport, but he also highly believed in physical conditioning! Not that Alf's physical conditioning drills were fun in any way, but I think most of us acknowledged that they were necessary in order for us to build strength and endurance. Luckily, I liked Alf a lot, and I seemed to click with basketball more than I had with any other organized sport. I actually loved the practices and the basketball drills, whereas I recall tending to dislike baseball, soccer and football practices and much preferred the games in those sports. As a group of players in Grade 8, we also loved to shoot hoops in our driveways with each other whenever possible, so our skills improved every year as we grew taller, heavier and stronger as we aged. We played for hours and hours on end, playing one-on-one, half-court and H_O_R_S_E, which I recall was a contest for foul shots. I think it was a good, formative year for all of us on that Grade 8 team.

Once we got to Grade 9, many of us who played on the Grade 8 team went on to play on the junior basketball team that was coached by Ed Terris. By that time, many of us on the team clearly had our sights on eventually making the senior team, and we collectively worked hard to continually improve. Many of us eventually made that team, and by the time we graduated, we had played five full years together. Several members of our group started attending private basketball camps in the summer at Conifer

Basketball Camp in Snoqualmie Pass as well as Birch Bay Basketball Camp with a variety of well-known coaches like Ray Thacker, Chuck Randall, Ernie McKie and Stan "The Man" Kirchman and various mini camps at SFU with John Kootnekoff, the SFU head coach, in attendance. During one of those camps at SFU, Rick Barry, who was a star NBA player at that time, put on a shooting drill for us that was unbelievable. He went around the top of the circle and made dozens of jump shots, one after another and never missed. Those were great learning experiences that were also a lot of fun for us young guys.

During junior basketball, Ed was not really into the physical conditioning quite like Alf but spent more time on actual shooting drills and numbered offensive plays. Ed was a great coach. He treated us with respect and would become an even better friend to many of us over the years. Both Alf and Ed, though very different in style, were huge contributors to the boys' basketball program in North Delta at that time.

In the fall of 1969, when I was just starting Grade 10, I was asked by Stan Stewardson, the head coach, to play on the senior boy's basketball team, along with Bruce Lowe, Don Clipperton and Stu Graham. That was quite an opportunity for us younger guys because we got to play with much older and more talented players during our practices, league games and tournaments, and that really helped all of us mature and improve as basketball players. Stan was an extremely competitive man, and he was focused on winning everything. Stan's conditioning was simple—you did everything you could possibly do to push yourself, even if it made you puke. This included running lines on the gym floor and running up and down the open wooden bleachers (which has probably since been banned by the United Nations as a form of cruel and unusual punishment). Stan was a very dedicated coach who put literally everything into this team, and it made you want to put in as much effort as possible to meet his objectives. This was expected of every player. One of Stan's preferred methods of physical training was his timed conditioning drills. Early in the year, he would post three minutes on the scoreboard, then every player would do his own conditioning "all out" for the three minutes, choosing any combination of a variety of drills like running lines, bleachers, sit-ups, push-ups, standing jumps, etc. As time went on, the timer went up in increments and by the end of the season, it was at ten minutes. There was no opportunity to conserve energy so you could last the ten minutes. It was intended to be "all out" or you would get loudly called out for lack of effort or commitment to the team, and no one wanted that to happen. This was a self-motivated style of physical conditioning that heavily relied on one's own drive to win and their dedication to becoming stronger, faster and better. Believe me, ten minutes of that drill was a killer!

In that very first year, Stan arranged a trip for the team to play in a tournament in Healdsburg, California against some very serious basketball teams from around the San Francisco Bay area. I think the original idea was cooked up by Stan to organize a home and away trip to our alleged sister city of Cloverdale, California because Stan lived in Cloverdale, British Columbia, which wasn't too far from North Delta. I'm not sure what happened to that concept, but we got into the tournament nevertheless. Healdsburg was an amazing mid-sized town that had palm trees lining the centre boulevard of the main street. We had our eyes opened at that tournament, not only due to the skill level down there, but also because of all the American regalia, including the school marching bands and big crowds. Even more impressive were the fancy uniforms with glimmering tear-away sweatpants and warm-up tops as well as the impressive pre-game drills. One team I remember was Oakland Tech. Every player on the team was black, except for one token white guy, and they could all stuff the ball in the pre-game warm-up drills. We lost the first game to Tamalpais, who eventually went on to win the tournament. We won one game and lost another, but it was a lifetime experience. During the trip, Stan paired me with Dave Coutu, our best player, to room with our local family in Healdsburg, and I suppose Stan wanted some of Dave's attributes and work ethic to rub off on me.

One other memory I have of that winter was going on a weekend hike up Hollyburn Mountain with Ken Manning, who was on the senior basketball team; and Brad Taylor; Steve Taylor; and maybe Ken Smith, who I will talk more about a bit later. The hike involved trudging up a rough, snowy trail from the Upper Levels Highway in North Vancouver to a little rustic log cabin that Ken's family owned up near the top of Hollyburn Mountain. We were all carrying food, clothing, ski boots, skis, poles, and some of the guys actually packed cases of beer! Believe me, we were loaded down for the near-vertical walk in the bush. Once we got to the cabin, we chopped firewood and got a fire started in the woodstove to warm up the place. The next day, we skied in the old Hollyburn ski area of wooded trails. This involved a rope tow that ran off the rear axle of an ancient tractor. It was a lot of fun and a great memory, but as this was only my second time skiing, my borrowed lace-up boots were very uncomfortable and the combination of my Head 210 cm skis (way too long for my experience) and cable bindings were hard to manage. Needless to say, none of us had proper winter gear, so we were all thoroughly soaked in our jeans by the end of the day. It felt great to be in the warm cabin that night after all that exertion. I recall having to go to a basketball practice on the day we hiked back down the mountain, and I clearly remember being yelled at by Stan a bit that day for a lack of physical commitment. (He certainly never learned that I was skiing that weekend!)

Our basketball team had a great year in 1969–70, going 8–0 in league games, 36–6 overall, placing second place in the Fraser Valley Championships and sixth overall in the BC Championships. Considering North Delta was barely ranked and didn't even make it into the Fraser Valley tournament the previous year, this was a tremendous achievement for us all. In addition, Dave Coutu won the MVP and Mike McNeill was on the first all-star team in the BC Championship. I remember hearing Stan say sometime after the end of that season that he had just set our sights too low.

The next year, we lost four players—Bob Tonsaker, Bill Chursky, Gary Robilliard and Richard Rudd—but gained four strong contenders from our junior team: John Buis, Rick Leblanc, Kevin Burt and Dan Schweers. At the onset of the season, we set the highest goal—to win the provincial championships—and we worked hard as a team to make sure we did not fall short. We breezed through the season with a 32–8 record overall, going 10–0 in league play, then went on to win the Fraser Valley Championship, partly because Don Clipperton (still in Grade 11) hit a layup on a pass from Dave Coutu with only 11 seconds to go against Abbotsford to win by one point in the semifinals. That game is still referred to as the "game of the decade" in the BC High School Boys Basketball Association 75th Year

Anniversary (2020) record book. Doesn't that make Donny's shot the "shot of the decade" for North Delta? We went on to win the BC Championship by winning four games straight, and even with everyone contributing, every game was a nail-biter. Don Clipperton was again the unsung hero in that tournament, shooting unconsciously and making most of his foul shots while taking pressure off Dave Coutu, who was always very well guarded. Dave deservedly won MVP for the BC Championships and the Fraser Valley Championships, with Mike McNeill and Stu Graham getting on the second all-star team at the BC Championship. This was the culmination of all the planning, dedication and hard work Stan had put in ever since we were all students in elementary school.

North Delta Provincial Champions 1971

Our Grade 12 year playing for Stan was another great year, but it was different without Dave Coutu, who had graduated and had been the best player in the province for the last three years. The core group that played together all the way through high school filled the gap as best as we could, and we had a fantastic season, nonetheless. Our overall record that year was 43–3, again going 10–0 in league play, winning the Fraser Valley Championship 3–0 and placing second in the BC Championship, losing to Centennial and Lars Hansen in the final game. Lars, who was about 6'11", killed us inside and none of us could stop him from running us down and going straight to the basket. The referees were definitely not going to foul out the premier player in the province that night. I am told that 8,700 people attended the Pacific Coliseum for that final game, and that is still the largest crowd to ever see a high school basketball game in BC.

For me, competitive basketball was highly motivating, much more so than scoring high marks in courses in high school. I generally did okay in school and got pretty good marks without too much effort, but later on, I didn't apply myself well because I wasn't properly motivated to do more. I suppose it might have been different if I had my heart set on becoming a lawyer or a doctor and had rich parents who were able to set enough money aside to pursue those dreams, but that certainly was not the case. Regardless, playing basketball on that team was a great life experience for me in high school, and I bonded strongly with my teammates and our coaches.

Many of the players on that team continued on to university to play basketball, with several of them getting full ride scholarships. However, as I mentioned earlier, I had no inclination or desire to go to university at that time, and I certainly wasn't highly pursued by universities offering scholarships either. Notwithstanding those facts, I did receive a written scholarship offer to play basketball at the Kingston Royal Military College in Ontario, which I thought about for a full one or two seconds. Can you imagine me in a military crewcut in 1972, playing basketball, taking calculus and physics and marching in the parade square, right at the height of the anti-Vietnam War movement? The main point of me mentioning this little-known fact has to do with something my dad told me years later in about 1979, when I was a superintendent in Cranbrook. That year, he visited me at the McPhee Bridge on his way through B.C., and he said that he was impressed with what I was doing at age 25, but to "Just think where you would be if you took engineering at the military college." (I mentioned earlier that we were not exactly showered with compliments growing up.)

After graduation, I played for several years in the Delta Men's Basketball League along with a number of my high-school teammates and our junior coach, Ed Terris. Ed became a very good friend over those years, and we always enjoyed a couple of cold ones after each game. I also golfed with Ed every Friday night at Peace Portal Golf Course during a summer in the mid-80's along with his buddies, Jack Newnham, who was a scratch golfer (man he was patient with us) and Laurie Lougheed, who was literally a retired brain surgeon. This is when I was just beginning to play golf regularly and take the game seriously, but Ed was always very supportive even though I often sliced balls into the treeline on the right side of the fairway. I think that summer is when I perfected my punchout shot because I got to practice it a lot on that course. Ed was extremely patient with me as well, resisting constant adjustments to my grip and swing, and just saying that it was a good swing, but I was just off a bit in the tempo. After every round, he and I met up at the Sundowner Pub, which was close to both his house and mine, and we would have a couple of cold pints and a great time, often running into a lot of people we knew. Unfortunately, Ed died of cancer one day before his birthday in 1997, and we lost a great guy and a great friend.

Starting around 1995, I organized a couple of annual team reunion golf tournaments and dinners for the basketball guys and coaches, which was always a great time. After Ed's death, Stan Stewardson had the idea of turning those reunion tournaments into an annual memorial golf tournament in Ed's name, and we held the first tournament in September 1998. Aside from honouring Ed and reuniting with the guys, Stan had another motivation for organizing the tournaments. He wanted to make them bigger to raise money for scholarships for students who wanted to go to university but could not afford it. The rest of us loved the idea, and we were away to the races. Stan was a natural and motivated fundraiser who had been involved in dozens of causes since we'd known him, including taking students to Expo 67; the World's Fair in Osaka, Japan; our team's trip to Healdsburg, California; and so on. Stan had the ideas about how to make money, and he had so much energy that in later years, I was worried about his health because of how hard Stan worked for the scholarship fund at the golf tournaments. We had putting contests, mulligan sales, silent auctions, live auctions, wine draws, 50/50 draws, raffle tickets—you name it, we had it. After running the tournament for 12 years, we had about $100,000 in GICs and had awarded over $30,000 worth of scholarships to deserving students.

Stan and I didn't see much of each other after graduation, primarily because he continued working hard with his coaching and then after a couple more years at North Delta, he became the head basketball coach at Simon Fraser University, all while I was starting my working career. However, he phoned me every year on my birthday just to say hello, just like he did to every single player who ever played for him, which was awfully nice of him. We eventually reconnected when the Ed Terris Memorial Golf Tournament started, and we worked closely on that charity for quite a few years. Stan and his wife, Heather, invited a few of the team members and their wives to annual dinners held every November at their house in New Westminster. Those were great evenings spent reminiscing and hearing stories from Stan, some that we may not had even heard before.

Among other things, Stan was a very gifted storyteller. Heather and Stan had a great life, and it was always so nice to share some time with them. On my 50th birthday, Stan arranged for a special birthday dinner for me at Gotham's Steakhouse in Vancouver, which was the newest, fanciest and by far, the most expensive restaurant in town. Stan took care of everything. Not only did he and Heather pick up and bring my wife, Gladys, and me home afterward, he paid for everything. Not that Stan could not afford to wine and dine like that (he had become very well off working so hard all those years) but I never thought that was Stan's style. Still, he went to that effort and expense for me, which was a very special gesture. Unfortunately, Stan died of a massive stroke one day before his birthday in 2017, and we all lost another great man, mentor and friend. The golf tournament still goes on every September in memorial to both Ed and Stan, but has now reverted to a simple reunion and a trust fund still pays out scholarships to worthy students every year.

Stan Stewardson

I was extremely fortunate to have great men like Ed McKenzie, Alf Clark, Ed Terris and Stan Stewardson in my early life in sports, and their influence proved very valuable throughout my life and career.

CHAPTER 5—
Kenworth Motor Trucks

IN THE SUMMER OF 1972, immediately following graduation, I applied for work at Kenworth Motor Trucks in Burnaby. Kenworth built custom-ordered industrial and highway trucks on an assembly line in a big industrial factory, and it was a great source of employment for a number of former North Delta High students. I seem to recall that Joanne Naismith's father was the general sales manager of the plant (Joanne was in the same grade as me), and that was my opportunity to get hired there. Many of my friends worked there including my girlfriend, Shawyn McKellar, and Al Northrup, Gary Webber, David Billman and, I think, my brother Bill for some period of time between attending UBC and his regular, extended winter trips to Mexico.

The proven method of getting a job at Kenworth was to show up at the front office at starting time (7:00 AM) with a résumé, wearing the appropriate work boots and clothing and with a packed lunch in case they needed someone to start that very day. If they didn't need you that day, you showed up again the next day at the same time and on and on. Well, I was hired on the third day and went to work in sub-assembly, which was off the main assembly line. Here, I assembled components that would eventually be installed onto the trucks on the line. I think I was getting paid $2.75/hour. My specific job was to assemble the fuel tanks with the fuel gauge sending unit, stairs, support brackets, etc. It was a great job! The first task was to determine what trucks were next in the line for assembly and get all the specs for those trucks, then you would go around the shop and collect the necessary parts and components and bring them back to your assigned assembly area where you performed the assembly.

At that time, Kenworth produced seven trucks per day, so you were required to complete the sub-assembly for those next seven trucks in an eight-hour day. By noon of the second day on the job, I had it figured out and I could very easily complete the required sub-assembly for seven trucks in less than a full shift. I think it was mid-afternoon on my second day when my foreman (Mr. Leatham, as I recall) approached me and enquired, "What in the hell do you think you're doing?" I had no idea what he was talking about when he explained to me in no uncertain terms that "They build seven trucks per day, and that's all!" That was a first for me—getting in shit for overproducing, especially on my second day!

Well, as enjoyable as that job was, especially being in a nice, warm and dry environment, (and except for the traffic going over the Port Mann bridge in the morning and evening), it did not last very long. I received a call one night from Russ Nustad, the superintendent of construction for the Delta School Board, and he offered me a job as a construction labourer in new school construction. In fact, he needed me to start that weekend as a watchman for one of the new schools being built (Burnsview Junior High) then report to a different school Monday morning for the labouring position. I gave my notice to Kenworth the next morning and completed my third day without incident or fanfare.

As well, around the time of graduation on June 27, 1972, I must have tired of blowing out the gas line on the Triumph because I purchased a beautiful blue 1965 Mustang from Broadway Motors on Kingsway in Burnaby for the price of $1150.00. This was the seventh car I had purchased in a year and ten days. The Mustang had a 289 cu in engine with a four-speed manual transmission and blue carpeting and leather bucket seats. The bill of sale indicated that, prior to finalizing the sale, the dealership would replace the clutch, adjust the lifters, fasten the muffler, and replace one tire for $8.00. Looking back, I wonder what the dealership actually did to make that sale? Maybe they added a slick oil additive to the engine to quiet the noisy (probably worn out) lifters, or maybe they put some fine sawdust or oatmeal in the transmission? Maybe some hairspray on the clutch plates? Apparently, all of these things can mask a problem for a short period of time. There were lots of tricks used at the sleazier car lots on Kingsway in those days. In any event, that car became a serious investment to me. I put new mag wheels and raised white letter tires on it, reupholstered the seats, as well as more than the occasional bodywork and paint job whenever the body got dinged up. I can distinctly remember three different fender-benders with that car, all of which required serious bodywork and re-painting to various degrees. The first one was when we were driving home from Vanloueven's Dance Hall, which was located on 200th Street near 8th Avenue in Langley. This was a large, privately-owned hall that allowed people to bring their own booze and snacks and partake in community dances. (These dances were always a lot of fun!) On the way home on that particular night, Shawyn was driving because I felt I was somewhat impaired and shouldn't drive. As we crossed a little narrow bridge over a little creek heading north on 200th Street, Al Northrup, a good friend of ours who was also at the dance, attempted to pass us at high speed in his beautiful 1966 Beaumont Sport Deluxe, but failed to pull out enough to clear us and rammed us from behind. Luckily, the only damage to the Mustang was to the left rear quarter panel, and Al was more than happy to pay for my body work without involving the police and insurance companies. The damage to Al's car wasn't too bad either, but it was a pain for both of us to get the repairs done and find paint to match.

The second incident happened when I was driving in Clear Creek Campground in Cultus Lake Provincial Park. All of our gang was camping, and I went looking for someone in the campsite. I made a turn in one of the lanes, quickly realized that it was the wrong lane and quickly stopped and put it in reverse only to realize that there was another car right behind me. It was embarrassing, but it happened so quickly. It certainly reinforced the habit of looking behind you and not taking anything for granted when backing up. This time the damage was relatively minor to the back of the car, but nevertheless, it required a new bumper, body work and the re-painting of at least the rear trunk and valance. I still have the invoice from Delta Auto Body for these repairs and it added up to $161.86 including 10 hours repair labour at $12.00 per hour and a new bumper was $31.30! (This bill equated to about one week's net pay!)

The third crash I had happened when I was driving up to Manning Park on a ski trip. I pulled a bit suddenly into the old Sumallo Lodge on the Hope Princeton Highway for breakfast and then realized that the entire parking lot was a sheet of ice, even though the highway was in great condition. I slid right into the lodge as I had no chance of stopping on that ice while trying to pump the brakes and steer out of it. The incident luckily resulted in no damage to the lodge or gas bar, but my front left quarter panel and headlight housing was toast. I think I may have had the entire car painted during these repairs due to issues matching the colour with the paint used during the various other repairs. However, I was too cheap to take it to Sandy Morita, who was well known for excellent paint jobs on muscle cars and hot rods, so I let Delta Auto Body do the paint.

The Mustang was a great car, but I poured a fair amount of money into it over the couple of years that I owned it. I don't recall a lot of serious mechanical work other than one time I asked my dad to drop it off at Busy Bee Brakes for a $88.00 brake special while I was at work, and when I picked it up, I think the bill was $588.00! So much for the specials offered by the brake and muffler chains. When I sold it in 1974, I seem to recall only getting $1600 for it, even though it was in much better condition than when I bought it. Regardless, I loved the car for the two years that I owned it, and we travelled all over the place in it.

CHAPTER 6—
Working for the Delta School Board

IN REFERENCE TO MY ABOVE-MENTIONED sudden career change from Kenworth Motor Trucks, I had applied for the labouring position with the Delta School Board along with my good school friend Don Clipperton before actually applying for work at Kenworth, so it was really something I had difficulty refusing when they eventually called. Russ Nustad was Don's neighbour down on River Road in Delta, and he was the superintendent of all construction with the school board, so he was a really good guy to know. Interestingly enough, Don and I used to play pool at Russ Nustad's house when we were in high school because that was the only pool table we knew of before we were old enough to get into Kennedy Pool Hall. There were several other reasons why this new opportunity was an improvement over working at Kenworth, such as better utilization of my past experience in construction and my recent rebuff for being overly productive. I really didn't think I would be very happy over the long term at Kenworth when I was so limited in productivity and challenge. Notwithstanding the fact that the commute within Delta was much easier and shorter than crossing the bridge every day, the School Board position also paid $3.76 per hour, which was considerably more than I made at Kenworth.

I should also mention at this point that the Municipality of Delta was booming with population growth in that era and there were literally a dozen or so new schools and new kindergarten classrooms being built at that time. The other unique thing about the Delta School Board was that it self-performed all of its own construction work and did not go to tender to general contractors. In essence, there was only one overall contractor for all school construction throughout the municipality, including excavation, formwork, concrete work, reinforcing steel, electrical, plumbing, masonry, stucco, and painting, so it was quite efficient and trades could be shifted around from one school to another as required.

The job as the watchman for the weekend wasn't very exciting except for a nice lunch that Shawyn brought me as it was very close to her house. Apparently, they had suffered some losses and vandalism at that school; however, I got through the weekend without incident. I was then directed by Mr. Nustad to head down on Monday morning to Hawthorne Elementary in Ladner and report to Superintendent Joe Bahry, a short, unhealthy-looking man with a deep red bulbous nose. My first assignment at that jobsite, as personally directed by Mr. Bahry, was to move a large truckload of lumber from where it had recently been rolled off the truck onto the ground to another location around the perimeter of the building foundation that was presently under construction. That was quite an effort, but I got the job done, however, not without some snarly looks and derogatory comments from my new boss that first day. At one point, when I was packing 12'-long 2 x 10s, Mr. Bahry inferred with a snarl that I should be capable of packing three boards instead of two, so from that point on, I accommodated Mr. Bahry's request.

I learned pretty quickly that first day that Mr. Bahry was an unhappy man. It appeared he did not like me, or likely any other young person, and he wasn't happy having a young, relatively inexperienced lad on the job. I was 18 at the time, had played virtually all sports through my life and had spent every summer since I was 15 working in construction, painting houses or on the end of a machete for a surveyor, so I wasn't actually a weakling, but I think Joe preferred older, crusty types, like Frenchy, that he could go to the bar with. (More on Frenchy later.)

The next day was very similar to the first day, meaning I was tasked with moving the lumber again. This time, the lumber pile relocation was apparently to allow a concrete pump truck access for an upcoming pour. Looking back on this incident now, I could possibly appreciate that this was just a test of resolve on my part or Joe's method to either break me or make me stronger, but ultimately, in my opinion, it was a needless form of worker misuse. Yet, no matter how ignorant I thought this guy was or the tasks that he assigned to me, I pushed on, and although I was certainly tired at the end of the day, I was not going to let this guy get the better of me. My memory is not crystal clear on this incident, but I am pretty sure that I moved that lumber pile once more before any other tasks were directed my way.

Once the lumber piles were relocated to everyone's satisfaction, I was given the job of compacting a small area of sand within the foundation using a gas-powered jumping jack. (Maybe it was the base of an elevator shaft?) Even I, at 18 years old, without any formal engineering training at that point of my life, soon realized that continuous compaction of dry sand was not increasing the soil density or load-bearing capacity of the area or improving the foundation condition whatsoever, but my instructions were quite clear: "Keep compacting that area until I tell you you're done!" Again, it was quite apparent that this ongoing task was intended to either make me quit or make me question the intelligence of the superintendent, so I did neither and continued to run that compactor, going though several tanks of gas in the process. Even adding water to the sand would have improved compaction, but there appeared to be no means or will on Mr. Bahry's part to accomplish that. He was simply running me through the "school of hard knocks" I guess, but it wasn't making my newly chosen career any fun, and the lunch and coffee breaks were pretty quiet for me as Joe had apparently hand selected the rest of the crew and no one would question the boss. Frenchy, who I mentioned earlier, was a greasy apprentice carpenter and one of Joe's right-hand men, and he clearly enjoyed watching me go through the rigours of Joe's gauntlet for new, younger employees. After a few days of listening to the conversations at coffee times and lunch, it became pretty obvious to me that Frenchy went to the same bar as Joe. I bet money it was either the Turf, Dell or the Flamingo in Whalley where they liked to hang out. (Much later, I determined it was the Ladner Hotel.)

After a week or two of this type of menial stupidity, I was told one afternoon by Joe to report to the electrician to help him for the rest of the day. My job was to assist him in carrying longer lengths of PVC electrical conduit across the jobsite where I would be on one end carrying four conduits, two per hand, and the electrician would be on the other end. While walking by my shiny blue '65 Mustang in the parking area, the electrician stopped and started asking me about the car. Within seconds, one of Joe's main guys (Guenter), came over and literally booted me in the ass, yelling for me to get back to work or I would be fired. The electrician was not happy about that, but he didn't really understand the ostensible probationary trial that I was working under at the time, so we carried on without further incident. The bottom line was that I was being tasked by the electrician, and he controlled my movement when he stopped, so I was certainly not responsible for any delay of the work. I must acknowledge that I had fantasies from that point forward about Guenter following me into the tool shack one day and turning quickly with a shovel in my hand and banging Guenter squarely into his square head. Actually, as a side story to this little incident, when I told my father what happened that day at work, he was pretty upset, and he wanted to know where Guenter lived. It just so happened that I knew where he lived, and it wasn't far from our house. My dad went looking for him that night, and to this day, I have no idea or recollection if he ever found him or confronted him.

I mentioned earlier that I got this job along with Don Clipperton through his neighbor. Don was working on a different school than I was and actually started a couple of weeks before me. I believe that Don started at Tsawwassen Jr. High School, but had been recently transferred for some reason down to Hawthorne, where I was working. Apparently, after receiving his last paycheque, Don had somehow become aware that the other labourers on his site were paid $4.76 per hour, rather than our $3.76. This didn't seem fair to Don, and I think he went back to his neighbour and complained that he was working just as hard, or even harder, than the rest of the regular labourers and deserved the full union rate. We were all members of the same union, Local 602 Construction Labourers, so it did not seem fair to me either, but I certainly was in no position to demand more money—I was just trying to get through the physical and mental abuse and keep my job! I think Don had just quit his job with the school board around the time of this showdown because he was hired by the Safeway Distribution Center, which was where he really wanted to work in order to follow his father's very successful career with Safeway.

Don actually remembers that he reported by telephone to me, rather than Joe Bahry, that he quit on July 11, 1972, and he started with Safeway on Wednesday, July 12.

Well, Mr. Nustad apparently went to Joe Bahry to see if I was working out okay and whether I was worth the extra dollar. Well, you can just imagine what Joe had to say about me. The next day, which was a Friday afternoon, Mr. Nustad came to my jobsite and approached me to tell me that they were going to have to let me go because I wasn't worth the extra money!! . . . Deep breath now . . . don't erupt too badly . . . easy boy . . . count to five . . . easy now . . . be respectful . . . then BOOM—all of my frustration, anger and resentment came flying out of me! I told Russ in no uncertain terms exactly what I thought of Joe Bahry and what he had me doing. I called him an alcoholic psychopath, and I told Russ that I was the hardest worker on the entire crew (statement of which I had no doubt was true due to the burdensome tasks and expectations that had been thrust upon me). I went literally nuts for a full minute and lost all inhibition, insight and abstract reasoning, but whatever I did or said that day was somehow effective, and although Russ told me to calm down, he got the message loud and clear. He simply asked me to report to Elgie Stevens at Pebble Hill School on Monday morning and that he would let Elgie decide if I was worth the wage or not.

Well, I was finally rid of Joe Bahry, but who knew what was to come in my next assignment? In those days, there were a lot of ignorant assholes in the construction world who believed strongly in the school of hard knocks, and who knew what this Elgie Stevens would be like? I was prepared for the worst, yet I needed the job and had to succeed. Failure was not an option at that point of my life. I reported to Pebble Hill School on Monday morning and introduced myself to Elgie Stevens. He was a tall man with reddish-blond hair; a huge, wide, sweeping mustache; and a deep voice with a bit of a Scandinavian accent. He immediately assigned me to the task of sanding the roughly spackled D-grade plywood walls of the kindergarten area with a sanding stick so that carpeting could be applied to the walls. I undertook that assignment with the same energy and attention that I'd used moving the piles of lumber and compacting the dry sand for hours on end.

I think it was either the second or third day of this physically demanding and sweaty work when Elgie approached me in a fatherly way and asked me, "What the hell are you trying prove?" As I began to explain my predicament and my trial period of acceptance based on the situation with Joe Bahry, Elgie calmly told me to forget about Joe Bahry and that he had nothing to do with me anymore. All Elgie expected of me was "a day's work for a day's pay." It was a great lesson and that important axiom has served me well to this day. I don't want to take too much literary licence here but during that conversation with Elgie, I think his body and facial language clearly showed that he had little respect for Joe Bahry and that he operated his project and his men in a much more intelligent and caring manner. Needless to say, I loved working with Elgie Stevens, and I learned a lot about being a good man and a leader from him. I loved his gruff, manly voice, his large mustache and his sensitive, caring side. I have the fondest memories 48 years later of that experience with Elgie, and I think of him often when I think of the decent, intelligent construction people I worked with over my many years. Oh yes, and I got the raise to $4.76 per hour!

Life at work was remarkably better after that conversation with Elgie, and never again was I ever worried that I wasn't worth the wage I was being paid. No one can operate at 100% every hour of every day, but I soon came to understand what a "day's work" was, and when weather conditions were inclement and difficult, I actually seemed to be able to outwork many of my co-workers who tended to knock it down a couple of gears in the challenging conditions. I seemed to be able to slog through it and get a decent day's work in. This could have been due in part to the fact that people's expectations were somewhat lower when working conditions deteriorated.

Once our work at Pebble Hill was complete, I was transferred to Tsawwassen Jr. High School, which was a fairly large new school that was partway through construction at that time. I got involved in the stripping of large wooden gang panels on the concrete gymnasium and auditorium. We would lower the panels to the ground with a crane and then we would clean them, make any necessary repairs and then oil them so they could be hoisted back into position for the next pour. Over the next couple of months, I worked with two very interesting and likable Italians named Carlos and Emelio. Emelio was a short, rotund man with a round, smiling face. He was from Northern Italy. Carlos, on the other hand, was a tall, muscular man with protruding cheek bones, eyebrows and chin. He was from southern Italy. Both these guys were hard workers and could speak English, but it was broken English and they had very strong accents. After getting to know these guys quickly, it was apparent they merely tolerated each other because Emelio thought that coming from the north, he was much more cultured than Carlos, who came from the south. There was probably some truth to this as Carlos did not seem as evolved as Emelio. Boy, they razzed each other constantly, and they razzed me to some degree as well, calling me "capusta", which I understood at the time to mean "cabbagehead". That was a great job, and I particularly loved the production that we achieved. The appreciation of production, as something to show for your efforts, had a significant effect on me and served me very well later in my construction career.

Most of the formwork stripping we did was performed from swing stages that were suspended from cables connected to heavy steel hooks that went over the top of 40'-high concrete walls. Every time we had to move the swing stage along, someone had to go to the top of the wall, which was only 6" wide after the corbel reveal was removed, and leap frog one of the hooks to reconnect the swing stage in the new position. The first time I had to do this was a bit scary for me due to the height, so I inched along on my bum while straddling the wall and cradling the hook. That is likely one of the days the Italians called me a "capusta". From that point on, it seemed that I had to overcome my fear of heights, and I worked on it. After a few days of practice, I was easily walking along the top of the wall carrying the hook and resetting it as required. As part of my self-training to overcome my fear of heights, I sometimes walked around the entire building on top of the wall to get another box of nails for the carpenters from the other side of the building, all without any fall protection whatsoever.

During this time, I worked for the labourer foreman, Parky Parkinson, and alongside labourers like Rusty Johnson, Brian Budd (who later became the first winner of the Superstars Competition[2] and played professional soccer for the Seattle Sounders), and carpenters Rick Wild, Roman

2 Superstars was the very first made-for-TV reality show based on individual athletic performance.

Ewasiuk, Hollywood John and Ron Nustad, who was Russ Nustad's son. I also got to know Tony Schultz, who was the superintendent of this large project, and many of the other skilled tradesmen working on the job. I particularly respected how Tony Schultz was able to stay calm while managing so many details of our large, complex project and how well he treated the men. Working with men like Elgie Stevens and Tony Schultz certainly made my job more rewarding and pleasurable and encouraged me to continue working in this field.

After my work at Pebble Hill for the painting crew (as a human sanding machine), they got me back on their crew after the major concrete work was completed at Tsawwassen Junior. Working with the painting crew around the district was a pretty good job. The painters were all very well-trained, professional journeymen tradesmen, and many of them had apprenticed in Europe or Britain. They taught me the proper methods and techniques for anything that I was doing, whether it was opening a new can of paint, cutting in an edge, rolling block walls or cleaning paint brushes. These men took me under their collective wings and utilized me quite efficiently. Although I did at least my share of the physical work, like sanding walls and wood cabinets, I was also provided the opportunity to do all aspects of the trade including painting walls, doors, trim and cabinets.

The most memorable character I worked with in the painting department was Denny Rogers. Denny was always laughing, telling stories or jokes and listening to country music on his ghetto blaster. The tunes of the day included Charley Pride, Merle Haggard, Johnny Cash and Waylon Jennings, so it was good old country music, not some of the newer whiny stuff that came out later. Denny and I quickly became good friends. One day, Denny and I were painting together in a new kindergarten and Denny was working on a step ladder with a can of paint in his hand. When he came off the ladder he stepped into a tray of paint, flipping it wildly in the air and splashing it all over the newly installed carpeting on the walls. As he was losing his balance, his other foot landed square into an open gallon of paint, which spilled all over the floor. All Denny could do was laugh out loud hysterically. It seemed to me that this incident might be something a guy could easily be fired for, but all Denny did was laugh and laugh! While he laughed the mishap off, I worried about the reaction our painting foreman, Sam Marlin, would have when he saw the mess and realized the resulting damage and delays to the work. Denny was an unbelievable character, and he continued painting for several years after that. The last I ever heard about Denny was that he became a fish and game manager somewhere in the Chilcotin, and unfortunately, I never saw him again.

Another good guy who I recall working with as a painter was Mike Senkiw. He and I were about the same age, but he was a fully trained and skilled painter. Mike had a big, frizzy afro-style hairstyle, hence why his nickname was Hippy Mike. That was better than what I recall his father's nickname was—Dirty Mike. Mike Senior was a plumber for the school board, and he always seemed to be covered from head to toe in grease and dirt. Hippy Mike, Bob Radaske (our lead-hand painter) and I would sometimes go down to Point Roberts for lunch on sunny summer Fridays, and we would sit on the beach and share the pleasure of a six pack of Pabst Blue Ribbon or Rainier beer while telling jokes or stories. On at least one occasion, perhaps on a particularly hot and sunny Friday afternoon, the beer tasted so good that we were a bit late back from lunch, and when our foreman cruised by our job to see how we were doing, all he saw were our brushes and rollers in their trays, waiting for our return. I think Bob had a little chat with Sam at some point later about our dereliction, but luckily, the repercussions did not flow down to my level of the totem pole.

Mike and Bob were funny guys, and we always had a good time working together. As I mentioned earlier, there was no shortage of work with so many new schools being built, so our productivity made our days fly by, and achieving good production as a crew felt great. Once, Mike and I went together to a concert at the Vancouver Coliseum where we saw Crosby, Stills, Nash and Young, and I thought that it was one of the best concerts I had ever seen. Mike lives in Penticton right now and is fully retired, and I do bump into him every now and then. We both enjoyed those days working together.

During my time with the Delta School Board, I worked in many different trades and did many things including general labouring, placing and vibrating concrete, stripping formwork, packing and tying rebar, working on the painting crew, installing steel studs, installing suspended ceilings, installing drywall, helping the bricklayers, helping the stucco crew (nothing fancy there –just shovelling into the mixer, pouring the mud into buckets and hoisting up scaffolding), working on the floor-laying crew (again, this was more packing linoleum and carpet rolls up stairways) and working as a glazier helper. I even helped the electricians pack conduit that one eventful day at Hawthorne Elementary. It was a great job, and I made some good money. I also met many great people along the way, but at times it was difficult, strenuous work. By that time, I was pretty sure that a job as a manager at a paint store wouldn't pay too much money, and the hands-on construction work I had been doing for the last few years was definitely not what I wanted to be doing for my entire life, but at that stage of my young life, what else could I do? Like many people at that age I suppose, that question began to weigh on me heavily.

In September of 1974, I took some holiday time off and went to Hawaii with Shawyn for one entire month! The flight there was with Air Club International and cost $128.00 USD per person for a return trip out of Seattle. (The good news is that the US exchange rate was $0.97 CAD to $1.00 USD in 1974.) We somehow rented a nice condo on the fifth floor only two blocks off Kalakaua Avenue in Oahu for $500.00 for the entire month. That was the first holiday of that type for me. Even though many of my friends and my brother Bill had been travelling regularly to Mexico on the cheap, I had not yet been there, as I was always too busy working.

We had a great time in Hawaii, and I just loved it there. The warmth, the fragrance, the friendly people, it was all great, and I never felt cold for the entire month! Friends from North Delta, Rod Neil and George Sobolik, were also there at that time, so we joined up with them the odd time. One day, Shawyn and I rented a topless Volkswagen Thing and went around the entire island (yes, even the off-road sections). We visited Sandy Beach, Sea Life Park, Hanauma Bay, the North Shore and Paradise Valley with all the trained tropical birds. One of our friends from high school, Donna Fonseca, was already staying in Waikiki, and she was dating a local Samoan guy named Haloti Maille, or Howard, as we called him, so we hung out with them quite a bit. Howard had quite a few other Samoan friends who we met, and one of those friends was an entertainer named Josiah who played guitar and sang in a lounge in the International Marketplace on Kalakaua Avenue. This was not far from our condo, so we saw him play there many times.

On one particular day, a few of us went out to Sandy Beach on the southeast shore and went body surfing. The beach has serious waves and was not really suitable most of the time for swimming, depending on the winds and tide. That day was great for body surfing, and we were all having a riot, until one wave decided to tumble my 175 lb. body headfirst straight into the sloping sandy shoreline. The impact sent a shock down my body as I tested the compressive capacity of my cervical spine—that was enough body surfing for me that day. Afterwards, our Samoan friends taught me that you need to position yourself so you can roll with the waves and prevent this deadly spearing action into the sand. It didn't turn out to be a serious injury or anything, but there was pretty steady nagging pain in my neck and shoulder, and I couldn't turn my head or sleep very well for a few days. Unfortunately, over-the-counter painkillers had no effect at all. After a day or so of pain, Harold offered to go to his doctor to get a prescription pain killer, like 292s, as neither Shawyn or I had any medical coverage in Hawaii. Well, this sounded like a great solution, so we went along with him to his doctor's office and sat in the waiting room while he went in with the doctor. Beforehand, I explained exactly how the accident happened and exactly where it hurt so he could get the best prognosis from the doctor. When he came back out to the waiting room, he was rubbing his neck, obviously playing the role very well, and he had a prescription in his hand. Once we got outside, he explained that the doctor decided to give him a cortisone shot in the neck to relieve the pain! Those shots are very painful and this quiet, meek guy took the shot for me so I could get the 292s. What a guy!

Me and the Samoans in Hawaii (I think Harold is the guy on the far right)

Another highlight of that trip was scuba diving at Hanauma Bay. Both Shawyn and I had taken scuba lessons earlier that year at Dive and Sea Sports in New Westminster and we had our scuba tickets, but diving at Hanauma Bay was completely different than any diving that we had done in the Vancouver area. The water was warm and beautifully clear, and the plentiful colourful fish and coral were unbelievable.[3]

Other adventures I really enjoyed around that time of my life were our many trips to the White Lake Ranch near Penticton. It turned out that Mark Sinclair's neighbour in North Delta married a guy named Bud Tower, who was part of an old ranching family in southern British Columbia. Bud had recently purchased the White Lake Ranch, which was composed of thousands of acres in deeded and grazing rights and stretched from one side of the valley to the other just outside of Penticton. Mark often took his two horses there in the summer and helped Bud with chores and electrical projects on the ranch. He and Bud invited a number of us young bucks up to the ranch, including Dan Neil, Vic Seder, Kerry Grozier and me. We had a great time riding horses, rounding up cattle, hunting by horseback in the

3 In later years, I continued to scuba dive, and my most favourite dive locations include the Lanai Caves and the Molokini Cliffs in Hawaii, as well as Cozumel and the Sea of Cortez in Mexico and the Great Barrier Reef off of Cairns, Australia.

fall, sitting by the campfire and eating really good, farm-cooked meals. In the winter, we snowmobiled around the property or used the ranch as a base to ski at Apex Mountain, which wasn't that far away. One fall day that I remember very clearly is when Dan Neil went out early one morning hunting on horseback and shot a deer. I wasn't really an experienced deer hunter, so I didn't bother getting up at 5:00 AM that morning, but Dan needed a hand getting the deer off the mountain, and when he came into the room Victor and I shared, he was covered up to his shoulders in blood, which was kind of a strange and frightful thing to see first thing in the morning.

Cowboy Days at White Lake Ranch

One particular ski trip that a group of us took around 1974 was also very memorable. Just a few days before Christmas, Mark Sinclair and I left North Delta with Dean Corbett in his truck and camper and skied at Manning Park for one day, then we carried on to Apex Mountain, where we camped overnight and then skied another full day. Afterwards, we enjoyed some après ski socializing at the old Gunbarrel Saloon where people tried to pound nails into big wood chopping blocks with a prospector's hammer. I have no idea how this didn't result in very serious eye injuries when you missed because, after all, this was a very active bar! Later that night, we travelled up to Silver Star Mountain and skied the next day there. The fourth day we skied at Tod Mountain just out of Kamloops and enjoyed the great light powder, the sunny skies and the old cowboy-style bars on that mountain. We camped out in the parking lot that night. The next day, we dragged ourselves home down the old Fraser Canyon Highway for Christmas Eve and spent some time physically recovering with our respective families. I need to point out that in those days, it was almost a competition to see how many runs we could get in in a day. There were certainly no long, luxurious lunches, taking the afternoon off or starting late—this was serious skiing. On Christmas Day, we spoke to each other again on the phone saying how much we enjoyed the trip, then the crazy idea was proposed that we do the exact same ski trip once more! Well, we left the next morning after enjoying Christmas with our families and made the complete circle tour one more time, but I believe we skipped Manning Park on that loop.

Taking a Break at Mt. Baker While Skiing with Rick Hansen

CHAPTER 7—
Turning 19

ABOUT THIS TIME IN MY work history, I turned 19 years old, which was when you could finally enter into liquor establishments in B.C., so we all went out to the Pillars Inn in Tsawwassen after work on my birthday to celebrate. Well, I think I ordered 19 beers (they were only 20 cents at that time, so the bill would have been only $3.80, and after all, I was making $4.76 per hour!) and the waiter asked to see my ID, which, for some strange reason, I did not have with me that day. So much for that party! That was really ironic because I had been going into pubs for quite some time without any incidents up until that special night.

One of our favourite things to do on the weekend evenings was to go to Madigan's nightclub in North Surrey where we saw now-famous rock bands like Trooper and Heart during their genesis. Our regular group usually included Vic Seder, Dan Neil, Brad Taylor, Kerry Grozier, and Mark Sinclair, among others. I loved that place, and we all especially loved Heart when they played there. We were all huge Led Zeppelin fans and Heart played a lot of Zeppelin and they played it extremely well and, in my opinion, sometimes better than Led Zeppelin themselves. (Roger Fischer was one great guitarist!) Not to say that we didn't see Led Zeppelin live every time they came to the Vancouver Coliseum, but Zeppelin was often inventive and played differently live than what we were used to hearing on the band's recorded albums, so we enjoyed Heart because they played their songs in a polished fashion, exactly the way Led Zeppelin did on its albums.[4]

Anyways, I digress. Back to Madigan's. Heart's road manager at that time was a guy named Don Beall. He went to North Delta High School and graduated one year ahead of us. We played various sports with Don growing up, so we often sat with Don at the band's table getting to know Ann Wilson and the rest of the band during the breaks in those early days. If Heart was playing at other nightclubs, we often followed them there. They frequently played at the Zodiac Cabaret in the Royal Towers in New Westminster and we were actually in attendance there on the first night that Nancy Wilson played with the band. That was a memorable evening! The band was completed that night, and they went on to superstardom with their two beautiful ladies who were both powerful musicians. While I, unfortunately, claim no specific memory of this happening, Vic and Mark, both vividly remember us all dancing with Ann and Nancy at Madigan's during the breaks on at least one occasion. (Who did you think "Magic Man" was really about??)[5]

I have many fond memories of watching and hanging out with Heart in nightclubs and going to see every Led Zeppelin concert possible, and many years later, in 2012, I watched Led Zeppelin receive the Kennedy Center Honors during the annual gala tribute on live television. A generation of all-star rockers were scheduled to play tribute to Led Zeppelin to celebrate their achievements in American music while the three remaining members of the band (Robert Plant, John Paul Jones and Jimmy Page) sat in the audience with the President of the United States and other dignitaries and celebrities. I wondered aloud, "Who would play the tribute to Led Zeppelin? Who could possibly do better than Heart?" Moments later, to my absolute emotional shock, out walked Ann and Nancy Wilson. Nancy played an acoustic version of Stairway to Heaven and Jason Bonham, son

---

4 If you view a YouTube video of John Bonham playing Moby Dick, just imagine Ricky down in the front row sharing a bottle of wine and a joint with the rest of the crowd. Those concerts were fantastic! (Outside of basketball season!)

5 Remember, this was when disco music was in its heyday. It was certainly not my kind of music and it very well may have me who invented the saying "Disco Sucks!"

of original drummer, John Bonham, joined on the drums! What a tribute that was! That performance remains one of the best of all the performances at the Kennedy Center Honors, and it brings back remarkably emotional memories for me every time I see it.

Another place we liked to meet often in the evenings after a long day's work was the Scottsdale Pub, which was located not far away from where I lived in North Delta. On any given night, we could count on anywhere between 6 and 20 people sitting in "our area" along the wall having a few beers. As most of us still lived at home, it was a nice place to get out and relax and see most of your friends for a few hours in the evening. This was an extended group of people from both Newton and Delta, and we became known as the NewDel Blues. We hung out a lot together, camping at Cultus Lake, Penticton, Hedley, Kalamalka Lake and Christina Lake; playing in a softball league; skiing in the winter and meeting regularly at either the Newton Inn, the Scottsdale Inn or the Delta Lion Pub. In later years, the Sundowner Pub became a very popular place for our group to meet.

On June 9, 1973, I purchased my first boat, a red and white 14' K&C fiberglass with a 50 HP Merc outboard and an accompanying boat trailer. This was a basic boat with no frills, but it was perfect at the time for water skiing (most of us still weighed 165 lb. back then) and a bit of fishing by Tsawwassen or Point Roberts. This boat was christened the HMCS Beaver and one of my friends (I think it was Jim McKenzie) painted a beaver on the bow. Our extended group of friends often congregated at Crescent Beach in the summer evenings, and we would take turns water-skiing until it was totally dark. The last ski of the day was always the best because the water was at its flattest.

Me With the Mustang and My First Boat – 1973

We would also ski a lot at Cultus and Skaha Lakes when we were on weekend camping trips. One night, we decided to go water skiing at 2:00 AM at Cultus Lake, and I towed around Andy Millar and Doug Tomkins tandem in the moonlight! It was so smooth; it was beautiful. I remember that I towed them towards the moonlight so I could see the water was clear of any floating obstructions, then I would turn and take the same route back. Our large group of party animals used to drive the Cultus Lake provincial staff crazy, particularly one individual who we had many friendly run-ins with. His nickname was Ricky the Ranger. Whether it was excessive noise, loud music, too many cars parked in a campsite or too large of a campfire, Ricky would show up in full uniform to address these concerns. One time, he charged into the campsite with a large fire extinguisher to put out our campfire because he judged it as dangerously oversized. Another day, after pretty well running out of patience with our fun-loving, but somewhat rowdy group, he suggested that we all pack up and head over to the private Vedder River Campground. Bob Vernon, who unquestionably possessed the fastest wit known to man at the time, quickly retorted, "'Cause we like it vedder up here!" This had 10–15 people laughing so hard they were rolling on the ground. Ricky the Ranger just left quietly.

One day that summer, Brad Taylor and I were out boating around Tsawwassen on a beautiful, calm, sunny day and we looked over at the Gulf Islands and decided to head over there, basically following the ferry route to Active Pass. I now know that distance to be about 15 nautical miles, based on nautical charts, but it was hard to determine an accurate distance looking across the Gulf of Georgia that day without any charts or navigational training. Considering the water was very calm and the boat had a top speed of about 35 MPH, it should have taken about a half hour. Well, we headed over at a good, steady speed, and we were probably there in less than one hour. Once we got close to one of the islands, we spotted a sailboat heading north, running parallel to the shore, and we decided to go over and ask them what island we were close to. We were told it was Mayne Island and that Active Pass was just north of where we were. The sailor must have thought that we were from some other planet, not knowing where we were out on the high seas.

The islands were absolutely beautiful with the rushing tidal flows along the shoreline and the colourful arbutus trees on the high rocky cliffs. Well, we made it to the Gulf Islands, but having heard nightmare stories about the rough water in Active Pass, I was not going anywhere near there that day. We headed back across the Gulf without incident, but we had to jump the ferry wake a few times to get some airtime and see how the boat handled. It was an exciting day, and it was the very beginning of my love of the Gulf Islands and cruising. We no doubt headed to the Scottsdale that night to tell our mates of our seagoing adventures.

Looking back on that initial seagoing voyage, and now many years later, having completed every course offered by the Canadian Power and Sail Squadron (except Celestial Navigation), I realize it may have been one of the most foolish things that I have ever done[6]. I certainly don't regret doing it, but I am very thankful that we didn't get into any trouble. Just to name a few of our shortcomings that day, we had no auxiliary engine, no tools, no spare parts, no spare gas, no nautical charts, no depth sounder, no emergency kit or flares, no bilge pump, no radio, no knowledge of navigational aids, no spare clothing, no emergency food or water, etc., etc.! To paraphrase an old saying, "There are things we know that we know. There are things that we know we don't know. But there are things we don't know we don't know." This saying certainly applied to our collective boating knowledge that day!

One other memorable event that occurred with that boat happened during a weekend at Skaha Lake in Penticton when we were all water-skiing and all of a sudden, the motor quit working. At the marina, it was determined that it looked like my reed valve had broken in the engine and it would need some fairly significant repairs. Well, as it was the middle of a weekend, I decided to load the boat onto its trailer and tow it back home after the weekend to have the work done down at the coast. However, due to the anticipated cost of the repairs, some of the people in our large group at the Beachcomber Campsite (mainly Paula West) thought they would assist me in defraying some of the repair costs by collecting beer bottles at the campsite and loading them into the boat. Well, a number of people became pretty focused, or shall I say, possessed, with this idea and

6 There is a fairly long list if I really want to think about this!

pretty soon the boat was chalk full of hundreds or perhaps thousands of beer bottles, to the point where the axles were starting to bow. The camp management eventually caught onto this activity and took some umbrage to this ad hoc fundraising effort, claiming part ownership of beer bottles within the campsite. Well, there was a bit of a standoff when we were leaving the next day, but I think a negotiated settlement was easily at hand as most of the beer had been consumed by our group anyway!

On one of those many evenings that the group convened at the Scottsdale Pub in the late summer of 1973, I was sitting talking to a good friend of mine, Colin Corbett, about the course he was taking at BCIT in civil and structural engineering. It sounded fantastic to me because it involved surveying, concrete technology, structural design, soil mechanics, timber design and all kinds of really interesting stuff. Considering what I had been doing for the last several years, it seemed to be the perfect step to get away from the physical end of the construction trades (hand excavation, sanding walls, moving piles of lumber, wheelbarrowing dirt, hoisting buckets of stucco, slashing lines for the instrument man, etc.). The more Colin described the course, the more I was convinced it was exactly where I needed to go! So much so that I headed down to BCIT the very next morning and talked to their admissions people for the Civil and Structural Engineering Technology Program. After explaining my related work experience, which BCIT valued highly in those days, I provided my high school transcripts, which, unfortunately, did not include a couple of general prerequisites for the course, specifically, Math 12 and Physics 11. The instructor told me that while I might very well be accepted into the course, I would likely not make it through the first year due to the steep learning curve that would be required without those two prerequisite courses. Considering that I had a huge struggle to get through Math 11 with my teacher Bessie Collins, who was not well regarded for class control and achieving full class focus and commitment to the course material, the thought of Math 12 after being out of school for a couple of years was daunting. Also, I never took Physics 11 because it was not required for a scholastic graduation, and I heard that it was a tough course taught by Mr. Sakiama. This was very disappointing to me after getting so excited about the program and how it could potentially get me out of the labouring market.

Well, after that, it looked like I had only one option to get into BCIT and that was going to college, taking the required courses and then reapplying once I had completed the courses. As best I can recall some 46 years later, it seems that I enrolled in Douglas College in Surrey in January of 1974 for the spring semester and then returned for another term in the spring of 1975. Initially, I took college courses equivalent to Math 12, which were Calculus and Physics 11, as well as one business writing class, which was also highly recommended to get into BCIT. I had quit working for the Delta School Board so I could put my full energy into my studies. There was a lot of pressure to succeed, and I felt like it was a "do or die" scenario.

We dove right into both the physics and math classes, and I tried like hell to keep up with the material, but I admit that it was very difficult. The business writing class was easy and took no effort at all compared to the other two courses. I had a great instructor named Mr. Hoffman for physics, but it was difficult to stay current, and I felt like I was sinking quickly compared to the rest of the, generally younger, students. After the first three weeks of classes in physics, Mr. Hoffman had us do a review test just to see where everyone was so far. I was near the bottom of the class. I think I got seven or eight out of 25 on that test. Later that day, after class, with this stark illustration of my poor grasp of the material so far, I met with him and discussed what I should do. Basically, he thought that I was just having difficulty with the new language of physics. All the new terms and references were so new to me and he suggested it might just take some time for me to get comfortable with them. We also discussed whether or not I should pull the pin and possibly acknowledge that physics wasn't my strong point, but he said in a matter-of-fact way that I would get 50% of my tuition back if I quit that day and if I stayed until the end of March, I would still only get a 50% refund. That made sense. Why not plough through for another month or so and see how I was doing at that point? I was certainly not a quitter, so I decided to press on.

From that day on, though, I began to dramatically improve my study habits for both math and physics. For physics I had to basically go back to Day 1 of the course and carefully re-read the text book and do all of the sample questions for each chapter, using a yellow highlighter to underline all critical concepts and equations that I should memorize. It was a rigorous step-by-step process, and I could not take one step forward until I fully understood where I was in the learning process. I made detailed notes of all my studies on 8-1/2" x 11" lined paper in a three-ring binder so I could review them at any time when I got stuck moving to the next steps of the course. Every available hour was used during the day and every evening—there was very little time for girlfriends or socializing anymore. It was all-out desperation at that point. Well, as you can imagine, once I made the decision to go back to Day 1 and re-review the course material, the class continued on with new material, and it was like they were building the fourth floor when I was still trying to complete the foundations, so I fell even further behind. However, as I doubled down with my improved efforts and study methods, I picked up the required knowledge and gained speed in the process. An even more important thing happened at that point was that not only did I start to catch on to the phraseology and concepts of physics, I started to really enjoy this new world. Physics applied to everything in the world, and I really enjoyed the new understanding of how "everything" worked. I actually began to love physics and enjoyed explaining how things worked to anyone who would listen. I found that explaining how things worked to someone else actually improved my own grasp of the physics I had learned so far. It wasn't long before I caught up to the class and was able to participate more in class sessions. I got appreciably more out of the classes after catching up, but I never stopped my review and study process throughout the entire course. As time went on and my improved understanding continued, I was much more engaged and conversant with some of the top students in the class. Needless to say, I certainly did not quit the class at the end of March. I went on to succeed in the course, eventually scoring the third-highest mark in all three Douglas College campuses for the subsequent first-year university physics course at the end of the following term, a feat that I can only attribute to my outright desperation, determination and fear of failure.

That summer, I worked as a private painting contractor, painting many new houses as well as repainting many existing homes. I worked for one house builder named Coolacre Investments that was owned by Barry

Coolins, who lived across the street from Shawyn in North Delta. I painted a few homes of his in North Delta as well as some rental houses owned by Nels Sinclair, Mark Sinclair's father. Mark was a good friend of mine from North Delta, and he worked at the Delta School Board as an electrician when I worked there. I also painted the Taylor's house and the Moran house on 92A Avenue in my neighbourhood and the Gagnon house across the street from Shawyn. After those were completed, Barry had another four or five new homes he was building near the freeway in Abbotsford that he wanted me to paint. It was getting busy, so I hired my brother Bill, as well as Dean Corbett, to paint with me. The minimal records I still have today indicate that I charged anywhere from $375.00 to $750.00 to paint a completely new house, both inside and outside, including the cost for all equipment and materials (keep in mind that a gallon of house paint was about $8.00 and our wages were probably based on $5.00 per hour). I rented sprayers for the exterior of some houses and that cut the labour time considerably, but the masking took quite a bit of time before we could even get started painting. Painting turned out to be a good way of getting work and making some money over those years. My skills started from painting with my dad and painting neighbourhood homes, then were honed by working with master painters, improving my efficiency. Unfortunately, this trade certainly did not appeal to me for the long term.

Near the end of that summer, I had an opportunity to be a deckhand on a small commercial fishing boat out of Ucluelet on the west coast of Vancouver Island. Having grown up in Annieville, I knew there were lots of fisherman in our community and there were lots of stories in that era about making big money on the high seas. A little extra money would be a welcome shot in the arm as I was heading back to school that winter. Well, with my crew finishing up the painting contracts, I drove my Camaro out to Ucluelet and met up with Ed Erickson, who hired me to go salmon fishing with him on his 35' boat called Blue Adriatic. He was an Annieville guy, and I had played men's league hockey with him for the past couple of years. Another boat that went out with us was owned by Bernard Remmen, another Annieville guy. Bernard's 42'-long wooden-hulled boat was named Viking Girl. Well, 4:30 AM came pretty early, and we started preparing to head out to sea. The winds were blowing all night, the marine weather forecast was not great and it was not going to improve according to the continuous VHF radio resounding about the docks, so most of the boats in the harbour were staying put. The halyards were ringing on the masts of all the sailboats in the harbour, and the wind was whistling through the rigging to a deafening roar. But our two-man flotilla had made up their minds to get to the fishing grounds first. The diesel engine was started and the aluminium hull vibrated from tip to stern with the throb of the engine. We made last-minute adjustments to the various cargo in the cabin as it was going to be a bumpy ride by the looks of things. As soon as we untied the boat, Ed took the helm and guided us out of the slip and away from the calm harbour.

As we left the confines of the protected harbour, Ed hung onto the steering wheel, constantly adjusting the direction to keep the boat steady into the oncoming waves. We immediately started to bounce with the surging tide and oncoming waves. My position in the cabin was standing beside the galley counter, holding on with both hands with every bounce of the boat. As we moved offshore, we headed into the full blast of the wind and the incoming breaking waves. The boat vibrated loudly from the roaring engine and lurched heavily as each wave hit and crashed right over the wheelhouse. I caught frying pans and other kitchen implements flying out of the sink and tried to keep everything in its place, to little avail. Forward visibility was near zero as each wave crashed over the windshield and washed across the cabin roof. Once we were on top of a wave, I could momentarily see our second boat beside us with Bernard at the helm, crashing into the sea with every wave as the wind sent the spray off in torrents. Yet, we pounded on and on. At that point, it wasn't making any sense to me at all—we were barely making any headway, so was it worth the effort? Were we just wasting fuel and perhaps even risking our safety, not to mention our comfort?

The going wasn't getting any easier and after a while I began to feel the queasiness coming on from being in an enclosed space. Try as I did to keep my eyes on the horizon, it was impossible with the bow rising steeply and then crashing into every wave on each decline and the restricted visibility caused by the sea coming up the windshield and over the cabin at every wave. Soon, I felt pretty seasick, and I tried to be close to the head.

Well, as anyone who has been seasick knows, an enclosed space below is probably the last place to be. Well, I emptied my guts, but the sensation did not improve one bit. My guts just wretched over and over, and I dry heaved loudly. Eventually, after literally hours of this, I began vomiting black coffee grounds and it didn't look very good for me, although at the time, I had no idea what was causing it. It wasn't any fun; I can assure you of that. After being at sea for about eight hours, the decision was made to turn back. I recall hearing later that we made it about two nautical miles in that period of time. It appeared that I was going into shock, so once we reached the shore, I was dropped off at the Tofino Hospital where I was admitted. I stayed there, in bed, for one week, recovering from blood loss from a bleeding ulcer. The only memory I have of that hospital visit was eating bland foods and some little Indian kids visiting me.

Once I was released, I was still feeling pretty wasted and somewhat disoriented. Because of this incident, I had to quit smoking Colts and stop drinking coffee and alcohol for six months in order to let my stomach heal. I made my way home and, to the best of my recollection, I returned to house painting in Abbotsford. I know that when I left for fishing, I was a bit backed up with work and had to hire Mike Senkiw, who I had painted with at the school board, to finish up some of the work on evenings and weekends.

Earlier that year, I sold the Mustang and purchased a 1967 Camaro with a 327 cu in V8 and automatic transmission. It was the same dark sky-blue colour as the Mustang, and thankfully, I never had to do any bodywork or painting on that car. On June 21, 1974, just in time for summer, I purchased my second boat, a 16' Hurston fiberglass runabout with a 75 HP Evinrude, for $1600.00. That boat had a little more power than the first boat, and therefore, was better at pulling skiers out of the deep water when slalom skiing. By this time, there were quite a few of the group who were becoming pretty good skiers: Rick Hansen, Al Northrup, Mark Sinclair, Colin Corbett and Dean Corbett, to name a few. In those days, it seemed to be the goal to slalom with hard turns, as hard as the old boat could manage, let's say, but it was the most fun to jump the wake on one ski and try to get as much air as possible. This led to some very significant and entertaining crashes when landing on the other side of the wake. That boat also made it to the Gulf Islands on several occasions, but I think I was slightly better equipped and a little more knowledgeable than I was on my maiden voyage over there.

One memorable story with that boat happened at Skaha Lake when we were camping for a long weekend. It was a very hot day, probably around 100°F and there were around 20–25 people from the gang on the beach in the powerboating area, so we could just ski right off the beach while watching the airshow being staged at the airport, just adjacent to our beach. The problem was that there were police foot patrols along the beach that day, so it was difficult to hide the fully stocked coolers of ice-cold beer that were critically necessary for a day on the beach like that. Therefore, it became necessary to store the coolers in the boat a few paces off the beach. Well, in the heat of the day, we were taking someone for a long ski out in the middle of the lake, when all of a sudden, something broke in the leg of the motor. We were stranded and out of power. We tried and tried to get the motor into gear without any success. We sat there for a while, waiting for someone to cruise by so we could get a tow into the marina where my car was parked. But no one came near, despite our efforts to wave someone down. It was smoking hot and all we could do was dive onto the lake to cool off, climb back onboard and crack another ice-cold beer. This process went on for what seemed like hours. Well, considerable time went by and finally, we waved another boat down with our distress signals, and they towed us to the marina.

We positioned the powerless boat beside the single boat ramp, and I waited my turn to back the Camaro down and winch the boat onto the trailer. It was a very busy weekend and the boat ramp was like a zoo! Eventually, I backed the car down and we got the boat on the trailer. Once secured in place, I went to start the car, but when I turned the key, nothing happened. I tried and tried to no avail. People were gathering to assist in any way possible as we were tying up the entire boat launch, but for some reason, we were all laughing so hard! I even got out a crescent wrench, laid underneath the car and tried shorting the starter out, as I thought it might be locked up. After a few more minutes of absolute buffoonery, I discovered that the car was not in PARK! I quickly, but stealthily, slipped the transmission into park and then fired up the engine and we got out of dodge before the Sherriff showed up. We sheepishly drove back to the Beachcomber Campsite to angry snarling from the rest of our beach gang. Not only did none of them care whether we were okay, they were more concerned about where the ice-cold beer was. Unfortunately, it appeared that the heat and extreme thirst got to our so-called friends, and they just assumed we had absconded with their beer on purpose.

One other memorable event that summer was spending one particular afternoon in the Penticton Inn after a day on the beach. Bob Vernon and I were introducing ourselves to a number of hefty BC Lions football players who were in town for their training camp and asking whether or not any of them liked to chug beer for 25 cents. Then we introduced Paula West, one of the regular girls in the NewDel Blues, who would be chugging against them. They chuckled and thought that it was ridiculous for a woman to challenge them to a beer chugging contest. Their clue should have been that Paula was a pretty large woman with a big swagger, a bit of a smirk and she was wearing blue jean overalls. Well, they readily agreed and the crowd started to gather. Soon a tray of beer was ordered, and one by one, Paula annihilated those beefy big men. Paula could drink a beer by the count of "one thousand one, one thousand . . ." and the glass was slammed down onto the table before you could say "two." Quarters were collected and we eventually ran out of takers. Back at the Beachcomber Campsite that night, we made a big communal pot of spaghetti, and we had a great party with Paula continuing to perform throughout the evening!

That was a great summer, with lots of camping, boating, water skiing and parties. I recall that I was away every weekend from May 24th to Labour Day. On one of those long weekends, we were camping in Penticton Tent & Trailer on Skaha Lake when a riot started on the main highway where it ran by our campsite. Picnic tables were burning and a huge crowd of drunks had gathered. I should point out that none of our people were involved in that – fortunately our group was lively but not malicious. At the time, I happened to be in town with Julius Lutar in his beautiful, white 1962 Corvette convertible, just cruising around town with the top down. We had a young lady sitting between us in the car who we had just met that night, when, all of a sudden, we realized we couldn't make it back to the campsite due to the riot. There was a battalion of police in full riot gear marching on the highway just east of the trouble, ready to go in and crush the drunken insurrection. As we approached them on our only known way to the campsite, they waved us over and told us to get out of the vehicle. As Julius got out, he dropped a baggy of pot behind his leg but the policeman noticed it immediately. They grabbed Julius, handcuffed him and took him away directly. Another police officer pointed at me and told me in no uncertain terms to "Get that car out of here!" I quickly started to respond that I had had a couple of drinks earlier and perhaps should not drive, but he didn't let me finish and with a loud voice told me to get out of there immediately. Okey-dokey, I guess I am leaving with the Corvette and the girl! Because we couldn't get back through the riot, all we could do was cruise away the night, just like in American Graffiti. Eventually, in the early morning hours, we drove all around Skaha Lake on the east side through OK Falls and made it back to the campsite from the south, well after all the trouble was cleared up. I went into the police station the next morning and picked up a sheepish Julius from jail.

Not surprisingly, due to the foregoing stories and adventures, and in spite of working steady for the entire summer and fall, I found that after tuition was paid, less the student loans and bursaries that I received, in order to get through the next year without income and still pay rent and utilities, I would need to sell not only my Camaro but also my boat. With the cost of insurance, gas and repairs, I needed to cash these vehicles in. I made the decision to take the bus and hitchhike to college. I was poor again, but I was focused on finishing college that spring and getting into BCIT in the fall.[7]

The second term at Douglas College was much less eventful than the first term. I had learned how to make notes, how to study and how to stay current with the class, so it was much easier and less stressful. I moved out of the family home that year as my parents were separating and getting a divorce. I found a decent basement suite on 82nd Avenue in North Delta to live in while I attended Douglas College and it cost $110.00 per month. It was a pretty basic place—instead of a stove it had a two-burner electric

---

7 I think there was also a fear of having any debt as well so whenever I was going to school, I was reluctant in carrying any debt other than a low interest student loan.

hot plate that had two temperatures: extremely hot or off! Living there on my own was the next very small step in adulthood. I'd lived in a suite over by 126th Street and 96th Avenue for a few months some time earlier when I was with the Delta School Board and loved that, but this time the rent was a burden as I was in school and not working.

Around that time, I was playing men's league hockey at the North Delta Recreation Centre with a bunch of friends from North Delta, and I was thoroughly enjoying it. I think we played Thursday and Sunday nights from 11:30 PM to 1:00 AM. Our team was called the Aercon Flyers, named for a sponsor. On the team were Mark Sarchet, who played goalie; Mark Sinclair; Rick Hanson; Vic Sedar; Dennis O'Brian; Leo Guichon; Al Northrup; Dan Neil; Rod Neil; Sonny Rattray; and Wayne Lock. In contrast to men's league basketball, which I had been playing on Monday nights in the Delta League ever since graduating, you could actually hit someone in hockey without getting a foul called, and you could avoid being fouled out of the game. I loved that aspect of the game. With all the padding, you could literally have these huge collisions and you wouldn't even feel it. Well, except for one collision that I remember very clearly. Ken Smith, who was a good friend of mine, played on another team and was a great hockey player. He was also 6'4" and 240 lb., virtually the same as Bob Dailey, who was a defenseman on the Vancouver Canucks at that time. For some strange reason, knowing what a nice guy Ken was, I decided to rub him out on the boards as he brought the puck up the left wing. I felt like I was hit by a Toyota truck. Talk about a yard sale! There was equipment everywhere, strewn 50' in every direction. Ever since, I have never tried to take on the big guys who could skate; there were other, safer ways to defend. At that time, I wasn't a great skater, along with a number of others on the team, but we loved the game. Rick Hanson wasn't the greatest skater either, but he could score. One game, he got a hat trick with one puck off his knee, one puck off his helmet and the third goal was a deflection off his athletic supporter cup!

One of the best beers you can possibly imagine is the one you have following a hockey game. After sweating under all that padding for one to two hours, it was one of life's greatest pleasures, even if it happened at 1:30 in the morning! We would have a maximum of two cold beers each in the change room as we got undressed and showered and then we were often starved, so we would head down to the 711 store to get a hoagie or something to get us through the night. On one particular night, I was with Mark Sinclair in his Z28 Camaro on a trip to the 711. We were inside making our purchase at the counter when a Delta policeman entered the store and asked who the car belonged to. Mark responded that it was his. The officer went on to tell Mark that it was illegal to leave a running vehicle unattended and asked for Mark's driver's licence. As he was reading the licence, he politely asked Mark, "How long have you lived in Delta?" Without missing a beat, Mark responded, "I have been living in Delta since you were back in the prairies slopping hogs!" I had no idea where this response came from or why Mark reacted this way, but I was concerned that there could be a slight delay or detour in me getting home to my warm bed that night. I do not recall if there was any further discussion between those two, and all I remember is that somehow Mark got his licence back and we got into the car and took off before the cop said anything further. (Or that's how I remember it anyways!)[8]

Once I was finished with Douglas College, I worked a short while with the City of Delta again as a rodman but soon left because I could make more money painting houses. When I gave my notice to the City of Delta, I phoned a good friend of mine, Dan Neil, who was between jobs and unemployed. I told Dan there would be an immediate opening in the survey department of the City of Delta. He seemed interested, so I drove over to his place, picked him up and drove him down to the municipal hall for an impromptu interview. On the 30-minute drive, I explained how basic surveying worked and introduced as many survey terms as he could

8 This was not yet the end of my hockey career as I eventually became an assistant coach on my son Derek's hockey team in Abbotsford. I remember really enjoying learning so much about the game from the head coach, a guy named Nate McReady. He was an extremely knowledgeable hockey guy, and it was a lot of fun for me to participate in all of the drills with the players. Also, it was remarkable to see the vast improvement in the skating, puck handling and shooting of all the members of the team, including myself. For example, Derek came in to the league barely able to skate, but after three years, he was one of the best skaters and goal scorers on the team. One problem was that we didn't have beer in the change room after those strenuous practices!

absorb so that the City would not think they were starting out at square one with Dan. Well, he got the job, loved surveying and continued on with the City of Delta as a technician in their engineering department for 35 years, ending up as the senior development technologist!

According to my account receipts from where I purchased all my paint (which, for some reason, I still have from Cloverdale Paints), I continued with some house painting at that time. Then, in August of that year, I returned to work for the Delta School Board on the new auditorium at North Delta High School. This building was identical to the auditorium I worked on at Tsawwassen Jr., stripping the wall forms, and I was now involved in the installation of Gyproc sheets on the stepped, sloping ceiling of this big building. Several of us worked on scaffolding that was set up from the sloping floor of the auditorium, and we hoisted the Gyproc sheets up and screwed them into the steel stud framework of the ceiling. Obviously, hoisting sheets up from the floor and holding them above your head all day was a rigorous and physically challenging job, but I remember the feeling of accomplishment I had as we had a lot to show for each day's work. Thankfully, however, that wasn't to be my life's work!

As I mentioned earlier, my parents separated in 1975. My mother retired for a second time from nursing and continued living a quiet and content life in North Delta with my sister Kathy. My father was promoted to shift supervisor at Shellburn Refinery in 1973 after starting at the bottom and working there for 22 years. He never had any post-secondary education other than various courses he attended as part of his training at the refinery, one of which included a week-long course on safety and fire fighting at the Texas A&M near Houston, Texas. Nevertheless, unlike many old salts who started at the bottom and worked their way up, he always had a great respect for the engineers he worked with. I think he had a great interest in solving problems at the refinery, and he would often draw isometric diagrams of pipes and mechanical things that reflected his interest in the science of the various systems he was involved with. I know that he regularly worked out solutions and improvements to those systems with staff engineers. I think he would have really liked to be an engineer himself. He often mentioned to me that he had great respect for BCIT as a technical school, and he liked the fine, upstanding gentlemen that he saw coming out of that school in Burnaby. (BCIT students all had short hair and wore ties in those days but that was abandoned a few years later, before I got there.) Perhaps I subliminally thought that attending civil and structural engineering at BCIT would make him proud of me, but in all honesty, my decision to go there was founded on my discussions with Colin Corbett a couple of years prior. I guess it was reaching that goal of shift supervisor, along with seeing many of his co-workers have heart attacks and strokes at an early age (and some of his friends actually dying) that made him decide to retire early and take his pension after 22 years. Within two years, he married his childhood sweetheart from Penticton and moved to cottage country north of Toronto. Several years later, he and his wife, Pat, moved to a beautiful, ocean-front home south of Courtenay on Vancouver Island. He remained there on a full pension until he died at the age of 89 on November 29, 2013.

CHAPTER 8— BCIT

IN SEPTEMBER OF 1975, I attended the Civil and Structural Engineering Technology Program at BCIT in Burnaby, and it was a damn good thing that I went to Douglas College and took the courses that I did. By that time, I was living in an apartment in New Westminster with my girlfriend, Annette Bengtsson, who was one of the NewDel Blues and working full time for BC Tel in Vancouver. The first term was an immediate onslaught from each and every instructor I had, and each one of them handed out assignments as if it was my only course! This was much different than my college workload. We had 33 hours per week of classroom instruction and a three-hour lab on Wednesday afternoons. The instructors were all great, except for one, who I will deal with later. The program was so overwhelming right from the first day, that I think it was designed to have about 30% of students drop out by Christmas. Regardless, I loved the course material and my previous courses in physics, math and business writing at Douglas College certainly helped. Ralph Englund was a great physics teacher, and I really enjoyed the civil engineering applications of vectors and the strength of materials courses that were taught by Tony Elston and Bob Starr. Timber Design was also a great course that I believe was taught by Tony Barren. Survey, draughting and communication courses were a breeze, particularly due to my previous exposure to those fields. Gus Anderson was the survey instructor, and he was a great teacher and a good guy, but I remember he always had a bit of antacid on his lips. I became pretty good friends with Randy Beattie, Ron Parker, Mark Filmer, Tim Murphy and Tom Abbuhl, so we generally sat together in class and often studied together. We all had our strengths and weaknesses, but Tom Abbuhl and Tim Murphy were brilliant and meticulous note takers, so they were extremely helpful during times when some of us were not catching on. Actually, Tom was such a great student that he became an instructor at BCIT a year after he completed his second year. Nothing came easy for the rest of us, but we persevered as a group and battled our way through the course material.

One course that proved to be very difficult for most of us in that first year was math, which was basic technical mathematics and later calculus, taught by Mr. Wardroper. Even though I had taken two different calculus courses in Douglas College, which should have made the course easy, in my opinion, Mr. Wardroper was the worst teacher I ever had in my life, including Bessie Collins, who taught me Math 11 at North Delta. I could literally learn more from talking to a brick wall than Wardroper! As I now recall, every time you attended a math class and turned in the required assignment from the previous class, regardless of the effort you put in or the correctness of your assignment, you got 1% of your final mark, so if you attended all his classes and turned in a dated blank sheet of paper each time, you got 25%. At the end of that first year, I met with the department Head for civil and structural engineering, Quintan Lake, because I was concerned that my math mark might cause me to fail the first term, even though I had achieved 86% on the rest of my courses. I was mad, and I told Mr. Lake that Wardroper had no business teaching in that faculty. It was well known that there were serious complaints about him that went back for several years. Even Colin Corbett and Mike Pritchard warned me that he was a terrible instructor and their class tried to get him removed years before. Well, in the end, I did pass that first year and, almost like a rite of passage, if you made it through the first year, you were expected to complete the two-year course, as the thinning process was mostly taken care of.

With the first year behind us, we began to get into the stride required to complete all projects and assignments, sort of like I had to do in the first term at Douglas College. We even spent some of our Wednesday afternoon study sessions at the Villa Hotel Beer Parlour where we could let off a little steam. Some of the other more enjoyable classes were Concrete Technology, Survey, Hydrology, Soil Mechanics and Structural Design. All of these courses had really good instructors and were very relevant to my planned future employment in the field of construction and engineering.

After that first year at BCIT, in the summer of 1976, I got a job with the City of Delta again, working as a surveyor on some drainage canal construction in the low-lying farmland in Delta. Rene Payer was my boss and a BCIT engineering grad, so we got along quite well. My job was pretty simple. All I had to do was keep the excavation online to the correct invert grade and correct side slopes. As we only had one excavator working, which was a Warner Swasey Gradall with a very good owner-operator, Harry Sullivan, it only took a few shots every couple of hours, so the job wasn't at all demanding or that fulfilling on its own. To ease the boredom, I began riding to the dumpsite and back with one of the dump truck drivers, Reg Merkley. He and I got along great, and he was soon teaching me how to drive and operate the dump truck, which was fun for me. It wasn't very long until, between the odd survey shot, Reg was sitting in the passenger seat and I was doing most of the driving in that 16-speed Eaton transmission. There was also a water truck on site that I operated when things got hot and dusty, although that definitely wasn't in my job description. One of the other dump truck drivers that summer was Ab Botkin, who was a fine old guy and apparently, as I recall, one of the original guys who started B&B Trucking in Ladner in the real old days. It was a good summer job, between working in the sunshine, driving dump trucks and doing a bit of survey.

The second year at BCIT went pretty smoothly, as I recall, and all of our little gang made it through with flying colours. Second-year courses, while they still included math and physics, were more practical-based, and the courses were very interesting and engaging. One memorable event at the end of that year occurred on our graduation night. We were celebrating that night and were pulled over by the police at 3:00 AM in downtown Vancouver after closing down the Ritz Hotel bar and then Sneaky Pete's nightclub on Hornby Street. Apparently, we caught the eye of a Vancouver City policeman when another car pulled up beside me on Hornby while I was driving Annette's green 1974 Capri that was equipped with a V6 four-speed. Well, the other driver wanted to see what the hot little Capri could do in a little drag race on the empty street, so I faked a start, then backed off immediately, while the other guy took off in a roar. I don't recall what happened to the other guy, but we were pulled over and the officer wanted all four or five of us out of the vehicle, standing in a line. After showing our IDs and explaining that we had just graduated BCIT and we were all heading off in different directions within the next week, he asked where I was headed. I told him I was headed to my home in New Westminster. He said, "Fine, as long as you head straight there, I am going to let you go. Drive safely!" Whew!! That was a lucky night and it certainly wouldn't be likely to happen these days. I think that was the last time that I ever saw any of those guys despite being so close over those two years.

CHAPTER 9—
Ministry of Transportation and Highways (M.O.T.H)

Campbell River

TOWARDS THE VERY END OF my final year at BCIT in 1977, I applied for an interview with the Bridge Branch of the Ministry of Transportation and Highways (M.O.T.H.) as they were conducting interviews at the school. My interviewer was Roy Buettner, who was number two in charge of all bridges in B.C., second to Len Johnson, who was the chief bridge engineer in the province. The interview with Roy was very relaxed and went very well, and he offered me a job in Campbell River starting on June 1, along with three other graduates, including John Jang, Tom Abbuhl (the really smart guy) and Ken Christianson (I think).

Well, off to Campbell River I went, only a few days after my final exams were complete! Essentially, over the next year, I worked for several senior project supervisors on a total of 22 bridge projects on a new Island Highway extension being built from Campbell River to Port Hardy. It was a perfect job for me because of my previous exposure to construction and survey, but more specifically, because of my engineering training at BCIT. My role involved a lot of highly accurate surveying for each of the new bridges, as well as payment quantity take-offs, concrete testing, piledriving records, soil compaction and density testing, reinforcing steel checks and draughting. I was also required to read and understand both the drawings and contract specifications.

This was a new frontier for me, and while it didn't necessarily come easily to me, I enjoyed the new challenges.

I first started working for Rocky Vanlerberg, who was very experienced in all the aspects of bridge construction, both academically and practically, because he had worked in that field for quite a few years. Rocky was in charge of several bridges, and I learned a lot from him as he was a strict taskmaster and fairly demanding of his men, which was good for me at that time. We worked well together and became lifelong friends.

The survey aspects of the job were much more involved than I had experienced in the past. Obviously, it was very important to be extremely accurate and to triple-check everything in survey for bridges so the steel or concrete girders fit properly when they showed up on site. In 1977, this technology was still quite primitive (compared to now), so we utilized calibrated steel chains for measuring distances, taking into account adjustments on each measurement for tension, temperature, slope and sag. Theodolites or transits, as they were called, were used for setting out lines as well as vertical and horizontal curves. The old "four-poster" transit with a plumb bob was used initially, but more frequently we used Wild T1A transits that incorporated a vertical optical sight to position the transit over the correct point on the ground. The T1As were much easier to set up because they were equipped with three levelling screws instead of four, and we became proficient in using them over this period of time. Every angle and every measurement was doubled over and over so any minor differences were split and the accuracy improved throughout the process.

Concrete testing was also a considerable component of the job. For each concrete pour, I was responsible for checking the slump of the fresh concrete, recording the temperature of the concrete, performing an air entrainment test, completing a concrete density test for each truck load and then recording all of the measurements in a log. Concrete cylinders were also cast for every pour to measure the compressive strength of the concrete after 28 days.

I had rented a beautiful apartment in town overlooking the marina and Quadra Island across the water, and sometime that summer,

Annette got a transfer with BC Tel from Burnaby to Campbell River. For awhile, all of us "BCIT types"[9] were all living together in that apartment, prior to Annette moving over. It was okay while it lasted and it helped pay the rent, but it was much nicer when we had the place to ourselves. We had a lot of visitors come over from the mainland, and we were often out in the boat cruising around or salmon fishing. One local guy I met worked in an auto-electric store and lived out near the water at Willow Point, just south of Campbell River. Cal Bence was an avid sports fisherman and loved getting out fishing whenever we could. We often met up for a seafood dinner when we were not fishing.

One of the different processes I learned on the bridge assignment was the science and art of setting bridge deck formwork. Without getting into too much detail, this practice is required because all bridge girders have positive camber (humped up in the middle of the span) and may be slightly different than each other. The bridge deck surface needs to be smooth on design grades both longitudinally and transversely after all the dead weight (concrete and rebar) has been placed on the deck and the initial dead load deflection has occurred. That generally results in thicker concrete over the girders at the piers and abutments, but the deck thickness must remain constant between the girders over the entire span of the bridge. To achieve this consistent deck thickness, the deck soffit forms were adjusted up and down at every form hanger support (roughly 4' centres) adjacent to the girder to create haunches, which is the vertical difference between the top of the girder at that point and the top of the soffit form adjacent to that point. Still with me? Once all of the haunches along all the girders are calculated and forms set accordingly, the bridge deck machine is used as a template to check and confirm there is a consistent and constant deck thickness across the entire bridge deck. This entire exercise is fairly straightforward for a bridge on a tangent, but there were examples, such as the Upper Elk Bridge near Sayward, where the bridge was

9 Some of the older Bridge Branch project supervisors coined this term "BCIT types" to include all of us new young employees that came from BCIT and it had a variety of connotations from good to bad depending on the individual using the term.

on a curve and tangent to curve transition with varying superelevation as well as highly skewed piers and abutments. Now, that got very interesting, and in some cases where it got extremely complicated, that science turned into an art form. I remember quite clearly that as Rocky and I set the haunches and soffit formwork on the Upper Elk Bridge, Manning's superintendent, Doug Kazakoff, was keeping himself busy suntanning on the roof of the office trailer, wearing only briefs and holding a reflective tanning shield around his golden-brown face. No wonder I wanted to become a superintendent—you could let the Ministry do all the complicated work for you while you made the big money! I will give Doug a lot of credit though. Some of Doug's greatest attributes were his organizational and motivational skills on site. When he organized the delivery of the bridge girders by barge from the mainland, the transportation of the girders to the various sites with concurrent erection of the girders using multiple cranes at each location was extremely smooth and efficient. Doug, who was about 6'4" and 240 lb. of muscle, also worked weekend evenings for a nightclub in Courtenay, but based on what superintendents were paid, I think Doug had that job for the additional exposure to the ladies. Doug was quite the character when he was wearing his big gold necklace and full-length fur coat!

Once the formwork for the bridge deck was all set and rebar placed, it was time to place the deck concrete. In those days, Manning Construction often used a fellow named Fred Dickhut to provide the deck machine and finish the concrete. Fred was very experienced at deck finishing, and he and his old Borges deck finisher had probably placed more bridge decks than all the other machines in the province at that time. Fred was an expert in what he did, and he knew the science and the art of placing beautiful bridge decks. He wore high gumboots on every jobsite and ran his deck machine with a 12" crescent wrench, making constant adjustments with the machine and banging his wrench on the machine when he needed someone's attention if they were falling behind or when the fresh concrete was a little low in one area. Fred would essentially run the deck pour and direct the contractor's troops accordingly. After any deck pour, Fred loved to have a few cocktails with the crew, and he was very popular with all of the Bridge Branch staff. I was fortunate to work on approximately seven or eight bridge decks that year, so I was somewhat apprenticed in that specialty work. There would be many more to come in the future.

After a while, I went to work for Harry Sandwith, who was an older gent with big paunch and a huge cigar hanging from his mouth at any time of the day or night. He had more hands-on field experience than, say, academic learning, but Harry was such a practical guy, and he got the job done very well. He also worked well with the new BCIT grads, and he was somewhat like a father figure to me. I sometimes drove with him in the mornings to the jobsite and then back in the evenings, and I really enjoyed Harry's company. At that time, I was back smoking Colt cigars, so we could literally smoke each other out in the government truck. We worked extremely well together and work was a lot of fun with Harry. Often, on the drive home, he would use the truck radio to contact Rocky and some of the other supervisors in the area to let them know we were stopping for a "coffee." Well, be assured that there was no coffee being served at the Quinsam Hotel in Campbell River! I would bet that the hotel had to order additional kegs of ice-cold beer when our group congregated after a hot day's work. It was a jovial bunch of men who not only got along very well with each other, but enjoyed and respected each other's work and company. Those guys loved their jobs, and their main topics of conversation after-hours were all about the finer points of bridge building. In fact, their wives referred to these ongoing, and often never-ending, technical conversations as "bridge building."

The Bridge Branch crew tended to stick together when we were out of town, and we socialized with each other regularly. Harry and his wife, Pearl; Rocky and his wife, Shelley; Rick Lewis; Ken Hunt; and Annette and I were often out celebrating someone's birthday. We had a lot of fun, but the main topic of our conversations was always "bridge building."

Celebrating my 24th Birthday Party with the Bridge Branch in Campbell River – Attendees included Rocky and Shelley Vanlerberg, Doug Kazakoff, Harry and Pearl Sandwith and Ken Hunt

I was eventually assigned, under Harry's supervision, to one particular project called Trout Creek Bridge, now renamed Mohun Creek Bridge, which is located just north of Campbell River. The bridge was being constructed by Kingston Construction and the superintendent's name was Clem Buettner, coincidently, he was the brother of Roy Buettner, who hired me. Clem had an extensive carpentry background, and he was very good at his trade. Whatever he may have lacked in formal engineering training, he certainly made up for in intelligence and practical experience. He was a dedicated bridge builder and had worked on many bridges and industrial projects over his career.

Clem and I got along famously and we became instant friends, which actually became a bit of a concern to some of the project supervisors because they were worried I might be less than objective in my dealings with Clem, as the contractor, and me acting as the bridge inspector. That did not concern me as I felt that I had appropriate ethics, and Clem never attempted to take advantage of that friendship. We often went for lunch or dinner together and Clem became a regular dinner guest at our apartment in town. Many evenings after work in the summer, we would be out salmon fishing in my boat in Johnstone Strait with a few beers, having the time of our lives. Catching nice fresh salmon was just a bonus to the evening. I do not recall if Clem had ever salmon fished before, but he certainly enjoyed those trips. I had recently purchased my latest boat from my dad when he moved to Ontario to get remarried. A few years earlier, he had borrowed my 16' Hurston one weekend to take some of his work buddies out fishing, and he was hooked. He bought his boat new one week later. This was now my third boat. It was a still-fairly-new green and white 17'6" K&C fiber-glass with a 140 HP Evinrude, by far my best boat yet and very suitable for fishing, cruising and water skiing.

While Clem agreed that my job as an owner's representative with the Ministry provided me with a lot of exposure to all kinds of construction, he thought I was risking becoming a lifer in the job with the decent pay and the nice government pension at retirement. I was still only 23 years old at the time, and I never really thought that this would be my last job, but I was certainly enjoying the variety of work and the people I was working with. My goal, however, was to eventually work for a contractor, and I was actually so convinced of that, that I made a $100 bet (keep in mind that $100 was a lot of money at that time as a case of beer might cost $6.00!) with Clem that I would be working for a contractor in less than two years spent with the Ministry. I went on to boldly predict that I would be a bridge superintendent within five years!

As we were finishing off the final touches of the Trout Creek Bridge, Len Johnson, the Chief Bridge Engineer, made a surprise visit and worked directly with me to redesign the river flow and bank protection for the bridge. This involved some site surveys, draughting sketches and cross-sections and we continued working on this design until 8:00 PM one Friday night. Once he was happy that we had finalized that design, he told me that I was to report to the Taghum River Bridge in Nelson, B.C. at 8:00 AM Monday morning! (Keep in mind that I got married while living in Campbell River, and Annette had a job with BC Tel there, and we had all our furniture in our apartment in town.) Apparently, that was the way of life with the Bridge Branch. They paid you a whopping per diem of $14.50 per day for living expenses and that didn't matter if you were home or living out of town somewhere. They owned you and you went where Len Johnson sent you, although they paid for moving expenses. Needless to say, I left for Nelson on that Sunday morning for the beginning of a brand-new adventure

Taghum Bridge - Nelson

When I arrived at Nelson that Sunday night, I checked into the Savoy Hotel, which turned out to be a great move as it was a really old historic building that also had a great, lively bar and restaurant to enjoy after work. The next morning, I showed up at the bridge site that was about six miles west of town and met my new boss, Don Evers. Don was a short, rugged guy, about retirement age, and he was wearing a khaki work outfit. I learned right away that Don was a very experienced construction guy as well as a structural engineer, and I liked him instantly. He showed me around the site, which was a new highway bridge crossing the Kootenay River and included four concrete piers and four pretty sizable abutments. This bridge was actually two completely separate continuous spans with an earth-filled causeway between them. This was an exciting project in a spectacularly beautiful location!

Don, who was semi-retired, was retained as the project engineer for this job by the consulting engineering company, Bush & Bohlman, which designed the bridge. Don lived about one hour away from the site in Kaslo.

He had very extensive bridge experience, recently as a manager with Dominion Bridge in Calgary and going as far back as the second world war building Bailey Bridges in Italy following the retreat of the German army. He had also worked many years with Larry Bush, one of the principals of Bush & Bohlman. I was Don's assistant and the only other representative of the owner on site, so it was going to be an interesting, hands-on project for me. Annette was still working for BC Tel in Campbell River, but eventually she was able to transfer to Nelson, and we rented a really nice, old house on Vancouver Avenue, up by the golf course.

Hansa Construction was the general contractor for the substructure of the bridge and consisted of Hans Mordhorst as owner, Erik Johnson as project manager and Art Lundeberg as superintendent. Hans was an extremely bright engineer who was not afraid to try new ideas and push the limits of engineering. Both Eric and Art were very experienced bridge and construction people and were also very innovative and hard-working. It was a pleasure to work with them on this project and although Hans has since passed away, I remain friends to this day with both Eric and Art. It was a very cooperative atmosphere on the site, and I think we all worked very well together.

My first task on this project was to perform the survey and layout for all of the abutments and piers, which was a pretty big undertaking by conventional survey means, considering the overall length of the bridge and the number of structural elements. Therefore, I came up with the idea of performing the survey using an electronic measuring device, or EDM. I had some limited experience in my survey class at BCIT using such an instrument and I knew they were becoming more common in the world, so I convinced Don that we should proceed that way. The EDM basically worked by sending a beam of infrared light to a receiving target and accurately measuring the distance to the target. I rented a machine called a Distomat that was made in Sweden and we were away to the races.

With Don assisting as assistant, I first ran a loop traverse with the transit for angles and the EDM for distances along the length of the railway bridge that ran generally parallel to the new bridge alignment. Next, I set off survey points at 90 degrees from each pier and abutment centreline along that baseline so those elements could be referenced from that baseline. At that time, intersecting lines could be surveyed to each element and survey points were then installed at the centreline of each of those elements, other than the two water piers, of course. Once the working points of all six of the land-based elements were accurately located, every distance along the bridge centreline was measured several times for increased accuracy, then cumulative distances were measured between all points along the centreline of the bridge. All measurements were recorded in a manual table, as this was long before Excel spreadsheets were in use. (For example, we measured Point A to Point B, then Point B to Point C; then measured Point A to Point C to confirm that was the same as adding the first two measurements. This exercise continued on and on for all points along the centreline so that we had accurate measurements (checked multiple times) between every two points on the project.) Once this entire exhaustive exercise was completed, we could calculate the potential error between any two points between the elements by using this table. It was quickly determined that any possible errors were extremely minimal (+/- 1/8" maximum between any two points), which was more than acceptable for this type of survey. When Don and I had finalized this exercise, he discussed the survey methods used and the findings with the design manager in the consultant's office. However, because of this new method of measuring distances in the survey, and the fact that EDMs had probably never been used by the Bridge Branch for bridge construction at that time, there was some implied doubt about the resulting accuracy of the survey. The consultant and several senior officials in the Ministry were contacted to discuss our survey results, but in the end, even with our fairly exhaustive exercise of cumulative cross measurements that confirmed the overall survey accuracy, the consensus was that we should re-survey the entire bridge using conventional survey techniques to confirm our survey. Those conventional methods involved the use of steel chains (basically a long survey measuring tape) along with the required chain grips, chain tensioners, sag corrections, chain thermometers for thermal corrections and vertical angle measurements with a transit and slope corrections for each and every measurement between points on the loop traverse. Needless to say, this was a lot of additional work to perform and I was upset, but Don took it in his natural stride and just took the attitude that "We will prove to them our survey is 100% correct!"

Over the next week or so, we completed the conventional re-survey and we proved our original survey points were good. Then, directly out of left field, the consultant suggested that we hire a local legal surveyor to double-check the overall layout, which Don dutifully ordered. They came and did their survey in a matter of a few days and signed off with a letter that stated all of our survey points were within 1"!!!! That was complete horseshit, as we knew all of our points were within 1/8" and we knew that the legal surveyor's methods were far less accurate than ours. Regardless, with the survey behind us, the contractor got underway and they were away to the races. Both Eric and Art were out of the gate quickly with work getting started on several fronts. There was also rock excavation required for one of the pier foundations and one of the abutments. Don was keenly involved with the drilling and blasting as he had had a blasting ticket for years as well as a blasting handbook, which we referred to constantly, so I was able to learn quite a bit about blasting from Don on that job.

The abutment construction was quite straightforward, but the pier construction was much more interesting. All of the piers had the same hammerhead cantilevered pier cap design, so Hansa built one set of heavy-duty timber forms that were supported by heavy-duty, custom-made steel support brackets and cantilevered each side off the single column. I clearly recall the forms were constructed as jobsite-built trusses, with multiple layers of diagonal sheeting and a million nails in order to carry the cantilevered loads, and they were designed by Hans himself. Once the form panels were used at one pier location, they were then stripped off with a crane, cleaned, lowered into the river and then floated by boat to another pier where they would be erected into position by another mobile crane.

There were two piers in the two river channels where the riverbed was solid rock. The pier design at these locations was very innovative and involved setting onto the riverbed a stay-in-place steel form that was 28' in diameter at the base and sloped up to a 10' diameter just above the waterline, so it looked like an upside-down funnel. The base of this form was outfitted with three fabricated adjustable height steel legs and drill anchor templates that were spaced evenly around the perimeter. The entire steel form was also completely outfitted with horizontal reinforcing steel rings and sloping rebar that was spaced off of the steel form to provide corrosion protection. Once a barge-mounted crane accurately positioned the steel form in its correct location on its three adjustable legs and set plumb, a barge-mounted drill drilled through the perimeter templates deep into the bedrock of the riverbed. Double corrosion protection Dywidag bars were then installed and grouted into the bedrock at those locations, thereby anchoring the form securely to the riverbed to withstand any tipping, uplift and shear forces. At this point, a well-blended rip rap was placed from a barge into the river all around the circumference of the base of the form to a height somewhat above the base of the form. Once the rock was in place, hundreds of cubic yards of ready-mix concrete was then tremied underwater into the bottom of the pier. Once the fresh concrete spread out to the rip rap around the perimeter of the form, it began to rise and fill the form, displacing water upwards as the level of the concrete rose. Eventually, once the concrete rose above river level and near the top of the sloping form, a prefabricated column of 10' diameter rebar was installed above that point and then the round steel column forms were connected to the stay-in-place steel form. At that point, conventional concrete placement continued until the height of the column was achieved. The concrete operation began in the morning and continued well into the night until it was completed.

A close call occurred late in the night of the pour as we were trying to install the prefabricated column rebar cage. When the heavy cage got hung up and wouldn't drop into position on its own weight with the crane's hoist line slack, several of us with sledge hammers hung onto the cage, hammering individual bars to find where it was binding and to try and get it to drop down. After attempting this for several minutes, all of a sudden, the cage dropped several feet, exploding the circular ties as it dropped and causing those on the cage to climb quickly up while it fell to avoid being skewered on the vertical dowels coming up from the pier or falling into the dark river. Luckily, no one was injured.

Later on, once the concrete had sufficiently cured, the column forms were stripped off and the conventional forming of the pier cap continued. This was a very unique style of pier construction, which I have never seen again, and now, due to stricter environmental considerations, I doubt this method of placing concrete directly into the river would ever be allowed again.

Among his many other talents, Don Evers was an experienced scuba diver, as was I, so we were able to perform certain underwater inspections together during the water piers' foundation construction. One task in particular was to inspect and observe the height and general density of the rip rap being placed around the perimeter of the steel form. Although nowadays a marine contractor would be able to see the underwater work by means of sonar mapping, at that time, our visual underwater observations and measurements were the best way to determine if the rock was placed adequately to retain the fresh concrete. I think on one other occasion we dove during one of the concrete pours to see whether there were any serious leaks in the rip rap allowing the concrete to flow through and into the flow of the river. I seem to recall that Don got me a pay increase for any diving time that we performed on that job.

Nelson was a great town to live in, and it was close to Christina Lake for summer weekends camping and water skiing. We occasionally travelled to Spokane for shopping and the odd game of golf. On one trip, Annette and I went with Art Lundeberg and his wife, Linda, which was a lot of fun. On several occasions we spent time with Erik, Art and all of the wives at the beach on Kootenay Lake where Erik rented a big trailer for the summer.

The rest of the work remaining on the substructure was completed by November, and I was transferred to Kelowna and Penticton to work with Arnold Talbot on about December 1, 1978. It was sad to leave Nelson, especially just before ski season as it was a great little town, but going to the Okanagan was also a great move.

Penticton

The posting to Penticton was a dream come true because it was my favourite place to be and had been for a long time. The warm, dry, sunny weather, the ponderosa pines, the beautiful lakes and clay bluffs fit me perfectly. Also, my father grew up there, and I still had my aunt, uncle and cousins living there, so it was a great opportunity to live and work there. Initially,

I worked with Arnold on the Ellison Overhead north of Kelowna, which was under construction at that time, so I lived at a motel in Kelowna for the first while. However, soon after, I was officially transferred to Penticton to oversee construction of three highway bridges on the new Highway 97 bypass under the supervision of Arnold. Wow, this was the goldmine! One of the four-lane bridges was located right on Skaha Beach where the Okanagan River enters Skaha Lake and where we had camped for many years. The other two bridges were very close by and crossed the Okanagan River at Green Mountain Road and Ellis Creek.

Annette had to transfer from Nelson to Penticton, and we rented a nice two-bedroom rancher on Duncan Avenue, very close to Main Street. It was only partially furnished, so I had to buy a bunch of new furniture to get the home set up appropriately. I was pleasantly surprised how inexpensive furniture actually was as I was buying a bed, couches, coffee tables, lamps, etc. It looked like I would be in Penticton for about a year, including the entire summer season that was approaching, and that was certainly something I was looking forward to.

Working with Arnold was a lot of fun, and there were a few of us "BCIT types" (Ted Peters, Tony Bennett, John Jang, myself and one other whose name escapes me) working with Arnold getting all of the initial survey and layout completed before full construction got underway. Arnold was about 48 or 49 years old at that time, but he looked 30 and he was always very fit, well tanned and handsome. He liked working with the young guys, and we often met after work in our orange government trucks for a couple of pints of beer and to listen to the many stories of his bridge building past. Arnold absolutely loved beer and he could stealthily out-drink any man I knew, possibly even more than Harry Sandwith, who was considerably larger than Arnold. On one particular day, Arnold came down to Penticton from Kelowna (where he lived at the time) and told us young guys that he was taking us for lunch. His instructions were to meet him at the Penticton Legion at 12:00 noon, sharp. Well, we dropped our survey tasks as instructed and arrived on time. We joined Arnie in having a few glasses of beer and there was lots of chatter, stories and laughter, then more glasses of beer and a hot dog at some point in the afternoon, which I guess Arnold considered "lunch". The stories continued and the noise level rose considerably. In the end, we did not get out of the Legion until about 8:00 PM and off all our orange trucks went in different directions to get back to our homes. Later, it seemed to me that this may have been some form of a test of character or rite of passage that we were put through by Arnold—maybe to see if we could make it or who was going to drop down comatose and puking from eight hours of constant swilling of beer? I wonder whether we passed the test? I know that we all made it safely home that night and showed up for work the next morning.

The winter in Penticton that year must have been the coldest in a long time because Skaha Lake froze solid its entire length and Okanagan Lake froze all the way to Trout Creek Point. I clearly remember skating on Skaha Lake, as many people did that year, and I know that cars were out on the lake as well! I had been working steadily for almost two years with the Ministry, and I was actually making more money than I was spending, so I decided to buy two motorcycles: a Honda 250 XL for me and a Honda 125XL for Annette. These were brand-new combination bikes that were good for both on- and off-road riding. I had loved motorcycles since I rode Darby Madakoro's 80 cc Suzuki in the empty lot on 112th Avenue in North Delta when I was about 10 or 11 years old. All of the kids would take turns doing a couple loops through the obstacle course that developed on that empty lot. I also loved riding mini bikes as a kid on Green Lake with Don Clipperton, Doug and Dan Schweers and Pat Rhola. So, Penticton, with its local hills and nearby mountains and logging roads seemed like a perfect place to ride these new bikes. Annette and I were required to get motorcycle licences, but it was pretty easy in those days. I put quite a few miles on that bike in Penticton, discovering backroads all over the area.

CHAPTER 10— Manning Construction

Lillooet Bridge

IN MARCH OF 1979, WHILE I was living in Penticton, I received a telephone call one evening from Ken Sleightholme, who was the chief estimator and general manager of Manning Construction in Vancouver, enquiring about my interest in working for Manning. Knowing that I really wanted to work for a contractor building bridge projects, rather than supervising on behalf of the Ministry, I had previously submitted resumés to what I thought were four reputable companies. Three were companies I had worked with in the Campbell River/Kelsey Bay area bridge projects: Kingston Construction, where Clem worked on Trout Creek Bridge; Manning Construction, that built Stowe Creek, Lower Elk and Upper Elk Bridges; and Dura Construction, that built two logging road overpasses. The fourth was Kenyon Contracting, which I worked with in Kelowna and Penticton. Ken wanted to hire me as an assistant superintendent on a bridge being built by Manning Construction over the Fraser River in Lillooet. Well, the call out of the blue, many months after submitting the resumés, surprised me. I was loving my job with the Ministry, taking on more and more responsibility, living in a great place like Penticton and even making more money, but this was a huge opportunity to step towards my long-range plan. I responded to Ken that I would have to think about it a bit, but I am pretty sure after discussing the pros and cons with Annette, I called Ken back within the hour and accepted the job offer.

I could barely believe it. We had just gotten nicely settled in Penticton and I was going to be on the move again, and of all places, to Lillooet! So much for the warm, dry summer that I was looking forward to on Skaha Lake. When I called Roy Buettner, the guy who hired me to work for the Ministry in the first place, and told him of my decision, he was surprisingly very supportive. In fact, he said that the time I spent with the Bridge Branch was great training to work for a contractor and he encouraged me to go on and become a bridge superintendent. Roy told me later that he worked for many years for a contractor before going to the Ministry, so he knew from experience where I hoped to be. I then talked to Arnold, who was a little disappointed, but he certainly understood my wishes to work for a contractor in the same field. He spoke very highly of Manning Construction as well as its owner, Dave Manning, who he had worked with on several bridges in the past. I was needed in Lillooet as soon as possible, so I handed off my government truck in Penticton, got myself to Vancouver where I met Ken Sleightholme, picked up a Manning company truck and headed up to Lillooet. Oh yes, and I won the $100 bet with Clem! I became an assistant superintendent within six weeks of working two full years with the Ministry.

In Lillooet, I reported to Hugh Brown, who was the superintendent. Hugh greeted me warmly. I think he was very happy to get some help. Hugh had worked for Manning Construction for about 25 years, and he had built numerous bridges and industrial projects all over the province, working directly for Dave Manning. I read fairly quickly that Hugh was under a great deal of pressure from Dave on this job and it was hoped I could help reduce that pressure by lightening Hugh's load.

The bridge itself was quite a long one, with four large piers and two abutments, and this contract was only for the construction of the substructure. The project was well underway when I arrived, but Manning was operating with a fairly small crew of workers who were essentially working on one pier at a time while Franki Canada, the piledriving subcontractor, installed foundation pilings at another location. I recall that Hugh's carpenter foreman was a German guy named Carl Feldhaus, a European-trained tradesman who enjoyed complex layouts of curves and stairs, etc. He would try and explain his intricate layout method to us with his carpenter's pencil and a cigarette package as he smoked constantly. There was some pretty fancy curved formwork on the soffits of the pier caps as I remember, so Carl's layout skills were well used.

Manning was also responsible for producing concrete aggregate and batching its own concrete from a gravel pit near town. John Findlay, a long-time Manning employee who was the father of a good friend of mine, Art Findlay (who was a NewDel Blue), was in charge of these pit operations. Manning was also working with Ben McCabe, a local Lillooet contractor for the batching and delivery of ready-mix concrete because he owned some concrete mixer trucks. It was thought that Ben, as nice a man as he was, was not too fond of bathing and his clothes were constantly dirty and covered in grease, therefore he got the nickname, "ring around the body". John Findlay was well known on the jobsite (and around Manning in general) as a pretty grumpy guy, and he either liked you or he didn't, but he knew how to make concrete aggregate and he had lots of tricks to do it. Initially, he really didn't have much time for a young pup BCIT type like myself, but over time, perhaps due to my keen interest in making spec concrete aggregate, I learned a lot from him and we even went for a beer every now and then. Even Hugh, with all of his experience and seniority, had some challenges dealing with John. Rather than tell John to do something, it worked better to suggest something and hope for a positive grunt of acceptance back from him. Another local character who worked for us from time to time with his D8 Cat was Keith Norton, who was a rugged bachelor and somewhat of a local legend in the area, so much so that the local coffee shop had a foot-long hot dog they called The Norton!

In the first couple of weeks, I was getting to know the site and the men, but I was mostly used as a gopher or a labourer. I don't think Hugh was really used to having an assistant superintendent[10], so it was like, "Go get a truckload of 2" x 10" rough planks at the sawmill" or "Go deliver two boxes of nails to the east abutment" and on and on. I could see right away that one of the problems on the project was that it was nine miles one way from one abutment to the other, which included driving through town, over the

10 And I have to admit I was certainly not sure what the role of an assistant superintendent should be other than following orders given to you by the superintendent.

old suspension bridge and down the highway until you got to the other end. If things weren't well thought out and planned, you could have people driving back and forth all day long. That planning and communication eventually improved.

In 1979, Lillooet was a bit of a rough and dirty town. There were lots of rough bars and a fair amount of serious drunkenness going on, particularly with the local unemployed and the Indigenous communities. After living in the Jay Gee Motel and the Four Pines Motel for a while, which were none too fancy in those days, I decided to purchase a fifth-wheel trailer so I could live on my own and cook for myself whenever I wanted to. I ended up buying a 33' Coachman fifth wheel that was in beautiful condition, along with a beautiful Chevy pickup truck that came with it. I set up my new home in a small trailer park looking over the Fraser about six miles east of town. It was nice, quiet and comfy out there.

Two of the piers on the bridge were located in the river, so they required work bridges to access them. Sheet pile cofferdams were installed to enclose the entire foundation, with excavation down to grade being performed with a crane with a clamshell bucket. Steel foundation piles were then driven inside the cofferdam and a deep tremie seal was placed before dewatering started. Once the initial dewatering was completed, the conventional forming and concrete placement for the footing, column and pier cap was undertaken. This was heavy construction for me at its finest. Franki Canada was the subcontractor responsible for the installation of the work bridges, sheet pile cofferdams, underwater excavation and foundation piling, but as I recall, Manning was responsible for the placement of the tremie concrete and the dewatering of the cofferdam, which was quite risky. Once the +/-10'-deep tremie seal was placed and cured underwater to gain strength, we turned on the big dewatering pumps inside the cofferdam and waited for the water level to recede. As the water level was reduced by pumping, the differential pressure from the outside to the inside of the cofferdam increased with large leaks of water occurring all around the cofferdam. We used oakum to try and fill the openings in the sheets and dumped cinders around the perimeter of the cofferdam to close the leaks as best as possible. As we began to make progress with the dewatering, the greater the differential pressure was, the tighter the sheet pile joints now became and eventually the incoming water was easily handled with a couple of smaller dewatering pumps at the base.

Once the water level was controlled at the base of the cofferdam, the footing could then be formed conventionally, footing rebar installed and then concrete poured by crane and bucket. Next, the column rebar, much of which was prefabbed ahead of time, was installed by crane. After completion of the column reinforcing steel, large pre-fabricated timber forms were installed and the pier column concrete was placed. It was on the day the column concrete placement was underway that Dave Manning arrived on the jobsite for a site visit. During a break, he asked Hugh, in front of me, how the young fellow (me) was working out. Hugh's response was very supportive of my work, and he told Dave that I was even a good concrete vibrator man to boot. (I had a little bit of vibrator experience from the Delta School Board, and I knew the proper techniques from the concrete technology course at BCIT and from my work with the Ministry). Well, Dave just about lost it! He tore a strip off Hugh, right in front of me, for using me as a labourer when I was hired to assist him with the management of the project. From that day forward, I was not used as a labourer any more, and I was able to start helping Hugh with more managerial tasks, like calculating payment quantities, surveying, working with Hugh to schedule the work, assisting however I could with the aggregate and concrete production and locating a suitable source of rip rap for the project.

After thoroughly scouring the countryside, we eventually located a suitable site down behind the Shell gas station where we could drill and blast rip rap, so an old air track and compressor were sent up from the company yard in Vancouver along with an old snuff chewing driller that Manning used from time to time. The quarry was very easy to develop, and we were blasting away in no time, with John Findlay and his 966 Cat loader digging and sorting the suitably sized rock that was required to line the banks of the Fraser River in the area of the bridge. One of my jobs during that process was to schedule the blasts, notify local homeowners and have traffic control in place during the blasts, just in case of fly rock. As some of the homes were in a very poor area, it was another eye-opener going to these homes from day to day to let them know of our scheduled blasts.

(It reminded me of visiting homes along the river when I was surveying in Delta.)

Dave Manning was a guy who always liked to utilize local labour and equipment wherever and whenever possible, particularly in out-of-the-way locations, but as a building trades contractor, it was often difficult to find those resources in a small town with all union members. Lillooet was a good example of that. We had been using an old guy named Warren Field who lived in the area and owned an old 25-ton conventional crane. Hugh had known Warren for years, and he was a good, qualified operator who kept his crane in pretty decent shape, so it was handy for a lot of the cranage, but he wasn't a member of the union. Well, somehow the Operating Engineers Union, Local 115, in Vancouver got wind of this after some time and made a visit to the site to investigate. Don Railton and Jim Lippert were the business agents for the union and they met with Hugh and I in their motel room where they laid the law down to us in no uncertain terms that this practice was going to stop and that they considered it a major infraction that would result in a fine or monetary assessment to Manning Construction. Unfortunately, in the end, Warren had to be let go, his crane demobilized, and Dave Manning agreed to make a substantial financial donation to the Local 115 Sports Fund. Those guys acted like thugs, and I never liked the Building Trades Unions too much after that. It made no sense to me to have to mobilize a crane and operator from Kamloops when there was a perfectly good one in Lillooet. No one would have minded paying the appropriate union dues to be able to utilize the local resources.

As our concrete piers were being completed for the substructure contract, Canron Inc., the subsequent contractor hired by the Ministry to install the steel superstructure, was on site delivering and assembling large sections of the trapezoidal steel girders. In addition, they were installing a large highline system to hoist and erect the girders across the entire span of about 1000'. This involved the installation of a very robust, but slender, lattice steel head tower 195' high on top of the west pier and a similar tail tower 120' high installed on its own concrete foundation on the centreline of the bridge just east of the east abutment. From each tower, a 3" diameter wire rope gut line was installed whereby a travelling hoist would run over the length of the gut line and enable it to raise and lower sections of steel up to 60 tons in weight, as required. The highline towers were guyed with wire rope to buried concrete anchors poured into the ground both longitudinally and transversely. The tail tower foundation, as well as the side guy foundations, were simple footings where concrete was poured into holes excavated into the cobbly soil with reinforcing steel and lengths of wire rope looped into the foundations to be attached to the guys. I was personally responsible for the construction of the highline foundations, using a couple of men from the bridge crew and working off drawings provided by Canron.

One day, I was down at the Shell station having lunch, when I heard someone say that the highline at the bridge collapsed. I knew that Canron was in the middle of erecting the tail tower that morning, so with my heart racing, I quickly left my lunch and drove the nine miles around to the east abutment to see what had happened. I was worried that one of the guy lines had ripped out of my foundations, causing the collapse, so I was pretty upset. When I got there, the ~100-ton crane was flipped onto its side with its boom mangled and the tail tower was laying crumpled on the ground. The crane operator, who I believe was Bill Boyle from Kamloops, was okay but shaken, and no one else in that area was hurt, so I checked the three foundations we had built and found them all intact—my worst fears alleviated. I found out shortly afterwards that one of the ironworkers on the east pier was hit with the gut line as it fell across the river and he suffered very serious injuries. That was basically it for Canron for a while as the tail tower had to be re-constructed and the entire gut line had to be replaced. I think we completed the remaining substructure before Canron was able to return and complete the steel superstructure. Many years later, due to insurance claims and various lawyers involved in that accident, I believe it was finally determined that the highline collapsed due to the crane assisting in the raising of the tail tower becoming boom bound, causing sideways forces on the tail tower that proved critical.

Once I was finished in Lillooet in December 1979, I returned to the Manning head office in Vancouver where I assisted on a couple of local jobs including one for Harry Hare on a bridge deck in IOCO and some estimating for Ken Sleightholme. At that time, Annette and I were renting a nice duplex on 96th Avenue in Surrey, which was close to all our friends and family, so it was an enjoyable departure from working in the field.

McPhee Bridge - Cranbrook

A couple of months later, early in the spring of 1980, Manning Construction was awarded a highway bridge project near Cranbrook called the McPhee Bridge, which was part of a new highway between the City of Cranbrook and the Cranbrook Airport. Dave Manning told me that I was going to be running the job as a superintendent! The project included two large concrete abutments and three concrete piers up to 148'10" tall. Just after learning we were the low bid in that tender, Dave and I took a drive to Cranbrook with all of the tender drawings and specifications to look over the site together. I knew that Dave was proud to be building the bridge because he had a fondness for the area and the bridge site was not far from where he'd lived with his family as a teenager. He had developed some clear ideas about how the project should be built, and he laid out these plans for me during that visit. I should mention that on that trip Dave insisted each of us share driving responsibilities evenly. His method was to drive exactly 100 miles then switch drivers. On one shift on the way home when Dave was driving, we were descending the big hill on Anarchist Mountain at a pretty good clip when I noticed Dave was nodding off and heading quickly towards a slow-moving truck. When I reacted, he slammed on the brakes, went into a full skid and just missed rear-ending the truck. Needless to say, it was time to switch drivers even though he had eight miles left on his shift!

After we returned to Vancouver, I started making final plans to mobilize to the site, making formwork drawings and calculations for the pier formwork and ordering form materials and supplies. Once we received the formal contract award and all of the insurance and bonding was in place, it was time to get started. Dave strongly believed in a fast start and an early completion. He arranged for an older carpenter foreman by the name of Slim Vold, who he had worked with in that area quite a few years before, to work with me. All the other employees were called out of the local carpenters', labourers' and operating engineers' union halls. So, there I was at age

25, running an out-of-town bridge project as a superintendent, three years ahead of my schedule with Clem! This was another perfect opportunity to live in my fifth-wheel trailer, so I towed it up to Cranbrook and set it up in a nice park-like setting in the local RV park.

The foundations for the bridge went in quite easily, particularly the river piers, which required dewatering, just as Dave predicted. He had coached me diligently on the principles of excavating the footing quickly, dewatering immediately and having the rebar and formwork prefabbed and ready to install without delay so the concrete could be placed as quickly as possible. These were great lessons and they paid off very well, both on that job and future jobs.

One issue that I had was the very long driving distance from one side of the river to the other, which I recall involved 12 miles of windy road, some of it being rough gravel. Well, I had seen the impact that had to the overall organization of the work at the Lillooet Bridge, so, after performing some cost analyses, I quickly decided to build a temporary bridge right at the bridge site, contrary to Dave's instructions and firm belief that this was unnecessary. I located a couple of fairly large used steel beams in Cranbrook that were just about the right length, then I performed some capacity calculations and found them to be just a bit light for fully loaded mixer trucks. However, rather than try to find different beams in the area, my idea was to reinforce the bottom flanges to increase their load capacity. That required adding a ½" plate and welding it full length, which worked beautifully. I asked my engineer friend Don Evers to double-check the design capacity and stamp the girders for me. The abutments for the temporary bridge were made of timber cribs filled with rocky material and the deck was installed with timber decking and bull rails. It all worked like a charm and was simple to construct and install. In no time at all, I had full loads of reinforcing steel, mixer trucks and our crew's trucks going back and forth over the bridge, saving miles and countless hours of additional travel. One other improvement we made on this project was the use of walkie talkies for communication, which made communication to the top of the piers much easier and saved a lot of driving back and forth across the river.

Temporary Work Bridge – McPhee Bridge

Goodbrand Construction, which was a relatively new company in B.C. at the time, was the general contractor building the new highway on both sides of the bridge, so they were also pretty keen to see the bridge installed. I was able to trade a few things with them for their use of the bridge, including the eventual removal of the bridge and approaches from the river, which they happily provided for me. Rick Hardy was the superintendent for Goodbrand and Kris Thorleifson was the project engineer. They both became good friends of mine along with Bernie Lofstrand, an equipment operator and Dave Winder, the crusher foreman. In fact, Kris, Bernie, Dave and I lived together in Cranbrook later that winter in a house owned by a lady we called Miss Kitty. She was a fantastic cook and a great hostess. We lived and ate like kings there and still banked money from our tax-free living-out-allowance cheques.

The McPhee Bridge was designed by Reid Crowther and Partners, a consultant, as opposed to the Bridge Branch, which was more typical in the past. As a result, the consultant provided an inspector and no Bridge Branch representatives were on the site during construction. The inspector's name was Pat Clancy. Pat was an easy-going guy, but quite a different type of person for this work. While not questioning for a moment his masculinity, Pat seemed slightly effeminate, much preferring to stay clean and perform office tasks than work in the field with the rough and tumble tradespeople. Pat had a little soul patch beard and was also known to wear an ascot on occasion. He often enjoyed a shot of a liqueur and perhaps a glass of sherry as opposed to drinking beer, mixed drinks or Tequila shooters, which were more common in our industry. As I recall, Pat had little or no past bridge construction experience; however, he and I worked closely together to make sure everything was being done correctly and efficiently. There was genuine trust between us from day one on the job. Generally, we surveyed and made various engineering quality checks together.

One day, as the foundations were completed and the river piers began to climb out of the river, Pat approached me, took me aside, and explained that he had a deathly fear of heights and could not go up the piers to do the quality checks and that he trusted me explicitly to perform all of the necessary checks on his behalf. That was interesting, to say the least—perhaps it was a good thing that I was a bridge inspector in my previous career. Ultimately, all the inspections on the piers in the air went swimmingly well—I never believed in cutting corners anyway.

The formwork for the three piers was designed with approximately 10" x 12" timber strongbacks running vertically with 2" x 10" horizontal joists using ¾" diameter high-strength steel coil rods and she-bolts, kind of an old-school method of forming for sure. The pier shape was basically two symmetrical columns with a curtain wall between them. What made things interesting was that the width of the columns decreased as the height increased, while the width of the curtain wall was constant. This made the construction of the inset panel form between the columns an ever-changing process from bottom to top, which required a lot of custom work for those insets. All formwork was completely prefabbed on the ground in panels about 22' in height, then hoisted into position using a 45-ton Linkbelt truck crane rented by the month from Grimwood Construction. Dick Robison was the local crane operator dispatched from the union hall, and he was a first-class guy and a first-class operator. He was all about safety. He knew his trade very well, and he was not afraid to operate at the engineered capacity of the crane, which we often had to do. The piers had five to seven lifts of about 22' each, and we poured concrete by crane and bucket every four days after we started the cycle. Inside the pier, we had hoppers and chutes to place the concrete.

McPhee Bridge – Centre Pier Built the Old-Fashioned Way

One day in the late fall, with the snow level descending lower on the mountains every day, we were on the final lift of the last pier, which was the highest of all three piers. I was doing the final survey checks from the ground to ensure that all the forms were in position and 100% plumb before ordering the concrete from the ready-mix plant in town. Once the survey was verified, I radioed up to Slim, the foreman, to confirm that everything was buttoned-up and that they were ready for the "mud". Slim said to go ahead and order, which I did by radio telephone from the top of the hill where I had coverage. After the order was placed, I returned to the pier and rode the headache ball up to the top to do my own (and Pat's) final cursory checks, and I immediately found a few problems. I could see pieces of copper tubing in the bottom of the form and then I noticed that the copper grout tubes that were soldered to the bridge bearing cans had been broken off. I quickly asked Slim what that was all about, and he admitted that the boys had some trouble fitting the cans into place in all the rebar with the grout tubes attached, so they broke the tubes off with a hammer. This presented a serious problem as there would be no way to pump grout into the cans once the bearings were installed. I told Slim that we could not proceed without the grout tubes and I would have to cancel the pour. I instructed him to have his men pull the cans back out so I could take them to a shop in town and have more flexible grout tubes re-installed. When I contacted the ready-mix plant, they reported that two eight-cubic yard loads of concrete had already left the plant and that they had no other places where those loads could be used. That was a lot of money wasted.

Once I returned to the office near the base of the pier, Slim came down and loaded the grout cans into my truck. I fired him on the spot, along with the two carpenters who broke the grout tubes off. As far as I was concerned, this was a careless and thoughtless act of sabotage, and they should have told me about the problem before taking the matter into their own hands and smashing the tubes off. I then left the site and telephoned the business agent for the carpenter's union in Cranbrook. I told him what had happened and that I had fired the foreman and two carpenters and that I needed them all replaced for the next morning. He said he would dispatch the men for me as requested. Well, the next morning, the repaired grout cans were ready to be installed, but the union hall sent out the same men I had fired the day before! I was so pissed off that I made a rare telephone call to Dave Manning to bring him up to date on the matter, and he in turn called the union hall. A while later, he called me back and said there was nothing he could do and told me to finish the pier with the men who were dispatched, as difficult as it would be. Boy, that was a tough one, but that's what we did. By this time, I began to hate what the Building Trades Unions stood for, how they operated and the power they had over privately owned companies.

One day, near the end of the bridge project, a guy by the name of Vern Dancy, who worked as chief estimator at the head office for Goodbrand Construction, approached me on site and told me that Goodbrand was very interested in getting into more bridge construction projects and they needed a young guy like me. Well, I kept this in my pocket for a while until the project was completed, but then after learning that Manning Construction was trimming its own staff after a very busy year spread out across the province, I decided to go work with Goodbrand after all.

CHAPTER 11—
Goodbrand Construction

Carnes Creek Bridge - Revelstoke

IN THE LATE SUMMER OF 1980, I started working for Goodbrand Construction as an assistant superintendent for Tom Beck on the Carnes Creek Bridge, 23 miles north of Revelstoke, B.C. This highway bridge became necessary because the new Revelstoke Dam was raising the water level by hundreds of feet in the valley and the entire highway to the Mica Dam had to be relocated far up the side of the mountains. As I was going to be there for several months, I hauled my fifth-wheel trailer up to a nice campsite just west of Revelstoke and lived there with Annette (who was pregnant with our first child) for part of the time.

The bridge substructure contract consisted of two large abutments blasted out of rock and three very large piers, the tallest pier being 250' high. That pier alone required over 5000 m^3 of concrete to complete. The bridge was designed by another design consultant, Hank Kolbeins, of Robertson, Kolbeins, Teevan, Gallagher (RKTG), so, again, no Bridge Branch supervision was provided. Tom Beck was a large, tall and rather dominating sort of guy with a big dark beard, but he had a great sense of humor most of the time. At times though, his emotions boiled over and his yellow Goodbrand hard hat was sent flying in a rage; however, that usually passed quickly and he would be laughing again in a short while! Tom was an extremely smart and shrewd guy, and I learned a lot from him on that very interesting bridge contract.

We had a stationary 250-ton Link Belt truck crane rented from Sterling Crane with ~240' of boom to service the centre pier in the canyon at a radius of 180' and the pier on the south side of the canyon. We used a smaller crane for the pier on the north side of the canyon. As this was a steep, mountainous site carved out of rock, the crane pad had to be constructed using timber log walls and buried logs as deadmen with 1-1/8" diameter wire rope tying it all together to gain some level real estate. Once the area was fully backfilled with rock, it became the temporary location for the truck crane and a work pad for the next several months to service two of the three bridge piers. Even prior to being able to construct the crane pad, a D6 Cat was yo-yo'd up and down the steep slope off a D8 Cat's winch line to clear most of the overburden above pier two. Then, a 225 Cat excavator worked its way down the steep slope on its own muck pile to the bottom of the canyon in order to excavate the centre pier foundation and assist with slope stabilization, rock anchors, shotcrete, etc., but there was no way it was coming back up that same route. One of the first tasks for the big crane was to lower an air trac down to the very bottom of the canyon where it drilled and blasted out the foundation for the main centre pier. A steel compressor line was installed down the steep mountain face to run the drill from a compressor parked up near the work platform. Cameron McIntosh was the rock foreman and therefore in charge of all blasting as well as anchor installation, shotcrete, etc. Steve Briscoe was the driller and another guy named John Upward took care of the installation and the grouting of all the tensioned rock anchors. Once all of the pier foundation work was completed down in the canyon, it was decided to lift the excavator out with the crane, as there was really no other practical way to retrieve it. To accomplish this, the boom was removed from the excavator to reduce the lifting weight because of the long radius required for the crane. It was a tense time for all of us as the multi-part crane line started twisting up as the hoe was hoisted up. Eventually, it made it to the top, the line was untwisted and the excavator reassembled without incident.

After the massive foundation was poured and the long pressure-grouted ground anchors installed, the pier column construction could commence. This steel column forming system was much more advanced than my earlier timber forming experience with Manning Construction on the Lillooet Bridge and the McPhee Bridge in Cranbrook. On this project, we used EFCO steel plate girder forms, which were self-aligned using adjustable support jacks, rather than the guy lines and tirfors that we used at the McPhee Bridge piers.

Another improved forming method we used on the piers was high-strength coil rod that extended outside of the forms instead of using pre-cut lengths of ready rod for use with she-bolts. The coil rod was placed in PVC sleeves made out of flexible waterline so the coil rods could be withdrawn when the forms were stripped and reused many times. This was much quicker and less labour intensive than using the old-fashioned she-bolts. Rather than a four-day cycle, which we experienced for each lift in Cranbrook, we cycled forms every three days on these bridge piers, with less labour and much greater accuracy and efficiency, even though the dimensions of the column constantly changed on all four sides every lift. Because I had somewhat limited responsibilities on that project, I enjoyed being able to work closely with the pier column crew each and every day to try and improve the schedule and man-hour efficiency for every lift. Barney Reynolds was one of the key young guys on that forming team who kept things moving. The prefabrication of rebar for each lift was also instrumental in achieving the three-day cycle. We had some great ironworkers on that project, including big Jack Regnier (I think that was his last name), but he was so big and round, I think he stayed on the ground and took care of the prefabrication.

Jack, Me and Cameron McIntosh

As it was more common in those days than nowadays, I believed it was important as a supervisor to have some familiarity with the various pieces of equipment used on your projects and, for me, this included learning how to run the Link Belt crane under the watchful eye of our crane operator, Ward Madge. Throughout the project, it was my responsibility to manage the pier crews, so I spent a lot of time on the catwalk of the crane with a walkie talkie as the only way up and down the piers was by using the crane; therefore, I was in constant communication with the crane operator and the crews on the piers. Over time, Ward taught me the basic use of the crane, and we both became very comfortable with my crane-operating skills. I eventually made full capacity lifts when removing or setting the large gang forms. That was a beautiful, smooth crane to operate and, although I have operated many cranes since, there is nothing to compare to that Link Belt.

We batched our own concrete just off the site in a portable batch plant, and a guy named Brian Koch was the manager of our aggregate production and concrete supply. We also had some very innovative ideas that transpired into methods of placing the pier concrete efficiently. One such method was the concept of transporting the ready-mix concrete for the main pier down a PVC sewer pipe that was installed and anchored down the face of the steep canyon into a large steel charging hopper. From the hopper, the concrete was discharged into a concrete bucket, as it was required, just adjacent to the pier. That system allowed the crane to literally hoist the concrete straight up to the top of the pier without a 180-degree swing or a lot of booming up and down, which would have been necessary if the concrete was conventionally placed out of the mixer truck at the top of the work platform. This saved a tremendous amount of time for each and every one of the 22 lifts that made up the centre pier, especially considering this involved about 5000 bucket loads.

Centre Pier Construction – Carnes Creek Bridge (Note the Concrete Delivery System Below Right)

Pier and Abutment Construction – Carnes Creek Bridge

In order to have water on site for mixing grout, green-cutting construction joints, washing out forms and curing concrete, etc., Tom devised a water-delivery system to collect water from a little sandbag dam in the creek way up behind the south abutment. This water then ran through a 1" diameter polyethylene hose all the way down to the canyon bottom, then all the way up to the top of the pier. In theory, as we all know, as long as the intake point is higher than the top of the pier, this should have all worked fine in an enclosed pipe system, except that when I initially tried to get the water running, even after thoroughly checking for obstructions in the line, I couldn't get water flowing through the pipe. After racking my brain for reasons it wasn't flowing, I had a thought, based on my fairly recent studies in hydraulics at BCIT, that perhaps the line friction and turbulence losses created by the small diameter pipe exceeded the pressure created by the head differential. I then performed a calculation from one of my engineering books based on the Bernoulli equation in fluid dynamics, which is $P = \frac{1}{2} p V^2 + pgh =$ constant, where P is the pressure, p is the density, V is the velocity, h is the elevation and g is gravitational acceleration. (Sorry if I have lost any of you readers with a less scientific mind than mine.) When I approached Tom with my theory, he burst out laughing and told me that I was nuts, it was just an airlock somewhere along the line. So, off a couple of us went, starting at the top and proving that we had flows every 100' by cutting the line, then re-splicing it, until we eventually had water all the way down to the bottom and all the way to the top of the pier. Tom actually demanded that this story be included in the book!

Labour efficiency was very important on that project, and in 1980, Goodbrand had quite an advanced costing system to accurately track man-hours and costs on any work item you wished. Tom was very instructional with costing and provided a lot of input to Al Rourke, the computer program designer in head office, about how the data should be collected, managed, displayed and utilized. This was the beginning of a strong passion I continue to have for tracking costs, improving labour and equipment efficiency and reducing related construction costs. This job-costing practice, if performed carefully, also provides a way to maintain records of achieved costs on a variety of work items and different projects, which proved very useful for future estimating of similar types of work.

The site inspector for RKTG was Jim Delaney. As I recall, Jim generally stayed out of the way but nevertheless, performed his quality control checks to confirm everything was built correctly to the specifications. He didn't appear to be an old hand in field construction, and I think Tom had little time or patience for him. On one occasion, Tom pulled me aside as we were heading into a monthly meeting with Jim in our site office trailer. All he said to me was, "You be the good guy, and I will be the bad guy." That's it, no other direction. So, the meeting got underway and not long after we got started into the discussion, we apparently hit the hot button topic where it appeared that Jim's opinion on this particular matter did not fully conform with what Tom thought it should be. Tom immediately stood up in a rage, threw his hardhat and stormed out of the meeting room, kicking his work boot right through an interior office wall on his way out! Jim looked white with shock and disbelief! He looked at me (I was also shocked) and asked what he possibly said to make Tom so angry? Now catching on to the game plan and taking the full opportunity to play the good guy, I explained in a calm and friendly manner our interpretation of the contract with respect to this particular payment item, and I concluded that we should be compensated for that work. Jim quickly acquiesced and could not have been more agreeable with me for the rest of that meeting. When I returned to the site and met with Tom, he howled with laughter at how well things worked out. It should not be overlooked that what Tom strongly believed was not a personal matter, it was a company matter, and he would do almost anything legal and ethical to make more money for the company.

The job profit on that job was approximately $800,000, which was a huge increase from the tendered markup, so I can only hope that Tom got a giant bonus that year because he deserved all of the credit for that fantastically successful project. One funny thing I remember about Tom was the way he handled some carpenters applying for work on the job and how he tried to assess their skills or experience. While Clem would have made the applicant build a sawhorse to measure his or her skill, speed and experience, Tom had a much simpler test: "Can you flip your hammer 360 degrees and catch it?" As this is a pretty easy challenge, most of them would respond confidently that they definitely could, but Tom would retort, "Too bad, we already have enough carpenters flipping their hammers around here, so we don't need you!"

Our first child, Derek, was born in the Revelstoke Hospital on October 2, 1980, while we were living in the 33' fifth-wheel trailer. That's life on the road for ya! Once the substructure contract was complete in the late fall, I returned to the head office in Aldergrove for a quick turnaround, but it was a short-lived breather. By that time, when I wasn't living out of town, we were living in a brand-new house that we'd bought earlier that year in Newton for $65,000.

Charlie Lake Dam - Fort St. John, BC

Once Carnes Creek was completed at the end of 1980, I was immediately shipped to Fort St. John to replace a young superintendent who was building a sheet pile cofferdam and concrete outlet structure on Charlie Lake just north of the town through the winter of 1980–1981. This project was paid for by Scurry Rainbow, an oil company that required a water supply in order to inject steam into their oil wells in the area (which I am guessing was the beginning of "fracking"). After the previous few projects involving challenging mountainous terrain, high concrete bridge piers and large volumes of concrete, this project, while not necessarily easy in any sense, was certainly less exciting. Unfortunately, this was one of those projects that would have been much nicer and more rewarding if I had started it from the beginning, but I was a junior guy with a new company and I went where I was told and did what I was told to do.

As this project was in the middle of winter, it involved a lot of heating and hoarding so everything within the sheet pile cofferdam had to be enclosed and heated constantly so the concrete would not freeze. Working this far north in the winter, keeping the diesel equipment running was another new challenge. In some cases, when it was extremely cold, we had to leave the mobile equipment running all night or it wouldn't start up in the morning. When things were shut off, we often had to get to the site early in the morning, cover the equipment with parachutes and get propane or diesel heaters going below the equipment to get them started. This certainly added another complicating factor to the work up there!

One old memory that comes to mind on this job involves the carpenters sloppily building the forms for one very narrow 3'-high weir wall in the structure and this formwork failed during the concrete placement. Unfortunately, it reminded me of a very tense day a long time ago while working for Kenley Adams. Now far from a catastrophe, this little failure was nevertheless embarrassing and it illustrated to me that vigilance is always required to be successful in this business. A superintendent was ultimately responsible for everything that went on a project, good or bad, and his or her reputation was on the line every single day. I don't believe that I was ever involved in another form failure from that day forward, but we came close in 1981 at the Agassiz Mountain Prison job in 1981, which we will talk about later.

One other fateful memory I have of this project was when John Lennon was murdered on December 8, 1980, by an insane gunman in New York City. That was truly a sad day, and I wrote this chapter of the story on December 8, 2020, 40 years to the day later!

Agassiz Mountain Prison

In 1981, I was directed by the newly appointed general manager at Goodbrand Construction, John Sinnema, to take over a construction project at Agassiz Mountain Prison. John had recently come to Goodbrand from Commonwealth Construction and apparently, he convinced Jim Goodbrand, the owner, that he could advance Goodbrand into a much bigger company, particularly in the industrial sector where Commonwealth had been a big player for many years. The existing prison was initially constructed to house, among others, the Sons of Freedom Doukhobors who were being arrested in large numbers in the Kootenays. This construction project involved upgrading the prison from a low-security facility to a medium-security facility. The project had already been underway for several months when I arrived, and as I recall, the previous superintendent, who was also formerly from Commonwealth Construction, had some problems, so management decided to make the change. This was the second time in my short career with Goodbrand that I was assigned to essentially clean up someone else's mess and finish a project. Unfortunately, this was too far to drive home every night to my home in Newton, so I relocated the fifth wheel to a park in Agassiz, but this job was far from a camping trip or a picnic.

This project involved a variety of significant tasks that included building a "tunnelling proof" perimeter concrete wall and roadway around the entire expanded prison, constructing a double row of new high-security fencing, adding four new guard towers and relocating several large existing buildings as well as all of the accompanying underground services. The interesting part of this job was that it was both within the existing prison and outside of the existing fence, so security was a constant issue. Guards were assigned to work with our crews at all times. Interestingly, the prison encouraged any employment of inmates that we could manage within the fence boundaries. One way I employed prison inmates was hiring them to perform test digs by hand shovel to determine exact locations for the various utilities on site before we commenced any machine excavation. These utilities included water lines, storm sewer lines, sanitary sewer lines, steam lines, electrical conduits and buried telephone and security lines. The prison had basically been built in an ad hoc manner whenever expansion was necessary over the last couple of decades, and there were virtually no as-built drawings and very limited knowledge of where lines were running underground, so this was a valuable service to avoid utility damage during construction. In some areas, we had four or five prisoners digging up an area up to 6' deep and discovering a multitude of lines. I know that one of those individuals on that excavation crew, who was a prodigious digger, was convicted of a double axe murder! As best as I can recall, the rate of pay for those individuals was the same as they made working in the prison, which was only a few dollars per day, so it was not only an effective use of labour, it was very cheap.

As I came into this job partway through it, I inherited the existing crew and one of the carpenter foremen was a guy who I'd worked with earlier at Revelstoke—with considerable disdain. He was a small guy who acted like a big shot because he had once worked on some of the big BC Hydro dam projects and he thought he was a big-time operator now, so we never heard the end of his experience there. Well, I figured I could handle him for one more job. One day, he approached me with his pronounced swagger and said that he would have a certain building foundation slab all ready to pour on Tuesday morning, but I reiterated my originally communicated plan to wait to pour the slab on Thursday morning so we could catch some other pours with the pump truck while it was there. His haughty response was, "Well I am not sure what my crew will be doing Wednesday because we will be all finished on Tuesday morning." My simple response was that he should just have it ready to pour at 9:00 AM Thursday.

Well, Thursday came along and we started pouring that slab foundation at 9:00 AM sharp, as planned. The far corner of the slab was just out of reach of the concrete pump, so we had decided earlier to wheelbarrow the first bit of concrete into that corner. I happened to be making a walking tour of the project during a regular monthly meeting with all the bigwigs, including the architect, consulting engineers, owner's representatives, prison officials and my new, fancy, suit-wearing, silver-tongued boss at Goodbrand at the exact time when the first couple of wheelbarrow loads of concrete were placed in that particular corner, with very concerning results! I noticed from a distance during my walk that each wheelbarrow load was causing the foundation formwork to deflect badly, which was not only shocking but infuriating to me. My new boss came over and whispered in my ear with his Dutch accent that I had a problem with this formwork. Without missing a step, I whispered back to him that it was, in fact, my inherited foreman from his ex-superintendent who had a problem. I then calmly walked over to my big-shot foreman and quietly told him to get those *#@&ing forms shored up immediately by any means necessary and not to, under any circumstances, stop or slow down the concrete pour, and I then returned to the walking tour. I was livid, but I didn't let that be known to anyone. The matter would be rectified, the pour would go on and the final lines and grade of the foundation would be perfect. It was just going to cause some people a lot of unnecessary stress in the short term!

Well, as I mentioned earlier, we were working within the confines of the prison for this part of the work and many prisoners were milling about watching the pour begin. I noticed one of the prisoners run to his room and come back with a well-used carpenter's belt, a sledge hammer and a handsaw, and he swung into immediate action, grabbing one of our circular saws and cutting stakes, pounding the stakes into the ground and installing bracing on the formwork to prevent any further buckling, all the while using a stringline for accuracy and straightness. I watched in amazement as this one-man crew outworked my big-shot foreman and my crew of inherited carpenters. He literally saved the day as well as my reputation with my walking tour of professionals and their ever-critical eyes. The pour did go on and the lines and grade of that foundation were perfect.

After the tour of the site was complete and all the white-hard-hatted professionals returned to their warm, cozy offices, I went over and had a discussion with that prisoner (whose name I learned was George Kinakin) who was still working on the concrete pour. George, who spoke with a fairly thick Doukhobor accent, quickly explained to me that he was an experienced carpenter and cabinet maker by trade (which was obvious to me from his work) and that he certainly saw the urgency of the situation and his need to assist. I told him that I personally appreciated what he had done and enquired whether he would like to continue working for us on that project. However, before accepting my offer, George felt it was necessary to explain to me why he was in prison in the first place. He retrieved a well-worn newspaper clipping from his shirt pocket and gave it to me to read. It appeared to be a statement by Queen Victoria in 1898 giving the Doukhobors the right to emigrate from Russia to Canada where they would be free to live by their own customs, beliefs and religion and have the right to refuse conscription. George went on to tell me that he was in prison for five years for burning down his house in protest of the treatment his community had received from the Canadian government, which was contrary to what they understood had been promised to them by the Queen a long time before. He was calm, rational, solemn and soft spoken with me but obviously still very passionate about the causes of his community and its people. He seemed to me to be a real nice guy and the kind of guy who would work well with others. He also seemed like a real earnest type of character who would most likely be able to do what he said he could do, unlike my present big-shot foreman. Then I got an epiphany: the idea of offering George the job of general foreman and ridding myself of my inherited foreman. The more I thought about it, the better it seemed. It made perfect sense to me as George was a little older and would likely do well as a foreman because he wouldn't be expected to break his back for eight or ten hours a day as a carpenter in the crew. George thought that was a great idea, and I knew he really looked forward to being busy every day with something he loved to do. George became my general foreman that day, and later that afternoon, I told my inherited foreman to pack up his things and move on!

George and I got along wonderfully, and he never once disappointed me. In fact, he excelled in everything he undertook. Considering his position, as best as I recall, he was allowed to work both inside and outside the prison fences as he was highly trusted and was obviously not considered a flight risk. When the project was finally complete, we were planning a windup dinner at a local restaurant and I pleaded with the prison staff right up to the warden to request that George could attend, but unfortunately, I had no success. That was a very sad day when George could not celebrate with us, even though he deserved it on many levels.

George showed me many of the woodworking projects that he built in the prison shop, including beautiful cedar chests with inlaid wood and other highly skilled woodworking projects. I learned from George's daughter Sara, when I spoke to her at length during the preparation of this book, that he actually built a couple of guitars while he was there. After I completed the project, George built me a beautiful, handmade wood family sign that I installed at the front of my home in Newton.

Many years later, while travelling in the Kootenays on a golfing trip, a few of us went to the Dam Inn, a pub in South Slocan, on the way home after a round of golf. I kind of recognized this guy sitting at a table, and when I approached him, he stood up and broke into a huge, wide, toothy smile because he recognized me and stood to warmly shake my hand! George was living in that area again, building cabinets and doing very well. It was such a great reunion that brought tears to both our eyes. That was sadly the last time that I ever saw George, but I certainly have some great memories of that proud and noble man.

> Postscript Note: In the summer of 2020, as I was travelling through the Kootenays on a motorcycle trip, I bought a book in a used book store in Nelson, B.C. that was written by Jon Lee Kootnekoff, who I knew from playing high school basketball many years before when he was the basketball coach at SFU. Shortly after reading the book, which dealt with, among other things, Jon Lee's heritage

as a Doukhobor, I tried locating Jon Lee and found that he was actually living in Penticton of all places—where I lived! After speaking to him a couple of times by telephone, he suggested that I read a book called *Terror in the Name of God* that was written by Simma Holt, to learn more about the Doukhobor religion and its history. I found the book online, purchased it and it turned out to be a very interesting, but quite likely slanted, history of the Sons of Freedom in Canada. The name George Kinakin was mentioned many times in the book. In follow-up conversations with Jon Lee, I learned that his girlfriend's last name was Kinakin, and I enquired whether she was possibly related to George Kinakin. Well, it turns out she was not related but she was a very close friend of George's daughter, Vera, and she gave me Vera's telephone number. Well, I spoke to Vera Kinakin by telephone at length on January 19, 2021 when I introduced myself to her and told her the story about me hiring George as my foreman at the Agassiz Mountain Prison many years before. Vera was a lovely woman to talk to, and she had all the time in the world for me. We shared many stories about George and she confirmed that the George Kinakin referred to in the Simma Holt book was definitely her father. In that book, George was described as a bomb expert and was allegedly implicated in many of the bombings in the early 1960's, including the grain elevator and the Anglican Church at Wynndel. According to that book, George developed the first timers for bombs that were constructed out of watches. On February 11, 2021, I spoke to George's other daughter, Sara, and we had a great, long conversation about George and the Doukhobor religion in general. I learned a great deal more from Sara. She recommended *Spirit Wrestlers - Doukhobors Strategies for Living* written by Koozma Tarasoff as another book I would be interested in reading, which tells the other side of the Doukhobor story.

Other than my sincere enjoyment working with George on that project, it was overall a very unpleasant job for me for a number of reasons. For starters, it was basically cleaning up someone else's mess with a strange, inherited crew, which is not often fun or rewarding. Working with the owner's inspector assigned to the project was also very difficult and extremely stressful because he was a supreme asshole. He was, unfortunately, the type of person who thought the best way to do his job was to make it as tough as he could on the contractor, which we have all seen before. This guy was a first-class weasel and did everything possible to trip up the contractor and its subcontractors. It was impossible to work with him, and I had to struggle to get through each day without a major fight with this guy. Working with the security systems and prisoner guards was also difficult and added another dimension to the lack of achievable efficiency on this project. Finally, working alongside the prisoners, no matter how safe one might feel, was an emotional challenge. In some ways, it was sad to see so many men confined. Other than the Doukhobors, a lot of the men were just simple, repeat criminals, sexual deviants or perhaps mentally challenged and that's how they got into trouble in the first place. It was really nice to get that job over with and move on!

Clearwater Town Bridge - Clearwater, BC

After a short stint helping out Frank Schultz, who was a senior superintendent with Goodbrand for the construction of the Maple Ridge Municipal Hall, a great opportunity came up for me. Steve McAlister, who was the contracts manager working in head office, was bidding the Clearwater Town Bridge along with Vern Dancy and he thought I would be a good guy to run that job based on my recent experience in bridge building.

Due to the perceived difficulties with the installation of sheet pile cofferdams and foundation pipe piling in the river due to the presence of boulders, no piledriving companies provided pricing for this project at the time of tender, so Goodbrand decided to price this very risky work themselves. I remember hearing that during the bid review, Jim Goodbrand asked Steve and Vern all about the sheet pile cofferdams in the river and how risky they were. Steve explained that they could be tricky, depending on the subsurface conditions and the fact that the bore logs indicated that boulders could be found in the river. Jim's response was to add another $200,000 to the bid to cover that risk, even though those complications were already evaluated. That was a significant amount of money in 1981. Nevertheless, Goodbrand was still low tenderer at a price of $2,580,380 when the bids were opened on May 11, 1981. Again, apparently due to the level of risk with the river piers, there were only two bidders on the project and the other bidder was Peter Kiewit and Sons with a bid of $2,998,958, a difference of over $400,000! Nevertheless, everyone was happy to get this project, especially me.

As I was named the superintendent for this project, I was able to select Clem Buettner as my general foreman. Clem happened to live in Birch Island, which was a 15-minute drive from the bridge site. Most of Clem's work had been located out of town, so this was a huge bonus for him to live at home for a change. Clem was excited to be working on the project but first we had to have a man to man talk to set things straight. Although we had kept in touch and remained good friends since Campbell River, we had not worked together since then and I was now a superintendent, no longer a junior, inexperienced inspector. Clem needed to understand that I was in charge of this job and he would be the general foreman, not the superintendent, as he had been in Campbell River. Clem readily understood and we shook hands with that understanding, most likely over a few Rusty Nails in the David Thompson lounge in Kamloops. This was a good meeting and it was vitally important to avoid any conflicts between us later in the job. (That may have been the same night that Clem proudly demonstrated how durable his Hamilton watch was by rapping it on the stone fireplace beside us, while pieces fell off of it and tinkled onto the table causing him to howl in uncontrollable laughter for some strange reason!) Shortly thereafter, off I went to that quaint little town up on the North Thompson River. One other twist of fate on that job was that Rocky Vanlerberg, my old boss from Campbell River, was assigned to be the project supervisor for the Ministry, so I knew it was going to be fun because Rocky was a good guy and we always got along great. I learned throughout that job that Rocky was very good at helping solve problems rather than creating problems, like some other supervisors and inspectors

I experienced. Rocky felt no need to knock you down every now and then just to show you he held the power, and he was just as proud as anyone of the innovation and good workmanship that we collectively achieved.

Me in My Office – Clearwater Town Bridge

By this time, Derek was almost one year old and Annette was not working, so I rented a nice, big house not far from the town of Clearwater and we leased out our house in Newton for one year. The rental house was located in Pumptown, which was an old company town built for employees of the Trans Mountain Pipeline that was built through the valley in the early 1950's. We got the manager's house, which had a nice, big, open layout with a loft for the master bedroom.

Soon after receiving the contract award from the Ministry of Highways, Steve McAlister received a call from Franki Canada, which was a foundation piling contractor based out of Eastern Canada. They had an interesting proposal to modify the foundation design of the piers to minimize the significant risks that were involved in installing the pipe piles and sheet pile cofferdams due to the presence of boulders known to be in the river. The alternative proposal included the installation of six 36" diameter x ½" wall thickness steel caissons per pier in lieu of the entire cofferdam and the foundation pipe piles. The installation of these caissons would be achieved by churn drilling, which was a common procedure used in Eastern Canada but had never been used on bridges in Western Canada. The process could drive right through solid rock or rock obstructions, so it entirely eliminated our earlier concerns. After considerable discussions with Franki, a completely revised pier design was proposed to the Ministry, and a credit was offered in the amount of $200,000 for further enticement. The Ministry seemed to appreciate this radical design change as it reduced its own exposure to claims if the original design could not be achieved, and it also saved another $250,000 for the Ministry-supplied pipe piles, which would no longer be needed. I remember Roy Buettner came up to the site very soon after we made our proposal and he, Clem and I had lunch at the Wells Gray Hotel where we collectively worked out a lot of the details for the new pier design on sketchpads. It was fantastic how everyone jumped on board with this completely new design concept and we were all extremely creative as a group to simplify the design quickly so we could proceed as soon as possible. We were definitely away to the races with this new pier design.

We got started on site with earthworks and installing the river embankments in August 1981. This included a significant quantity of rip rap, which we drilled and blasted in a pit only a couple of miles away up off Camp 2 Road. Cameron McIntosh got that work set up for me with a company air track and compressor as well as some preliminary blast designs. Steve Briscoe was again the driller and blaster and that work went very well. Franki soon started building the temporary work bridges to each of the river piers and began driving the caissons for the west pier. It turned out that this process of churn drilling was very crude but still effective for this particular application. It was basically a huge steel star bit weighing about 12,000 lb. that was picked up with the crane running at full RPM and repetitively dropped inside the pipe caisson to crush whatever was in its path. Typical advancement, even in rock, would be in the range of 12" per hour, so it was a slow but relatively steady progress. The chipped rock material was absorbed into suspension of the liquid within the caisson by adding bentonite in the pile and that liquid slurry was removed from the caisson by using a big suction tool on regular intervals then it was loaded into sealed dump trucks and disposed off-site. The churn drill excavation advanced ahead of the bottom of the steel caisson so the end of the pipe was left undamaged and the pipe was regularly tapped down close to the bottom of excavation then the process would continue.

Churn Drilling Piles – Clearwater Town Bridge

Until the east embankment and west pier were completed, there was really no carpenter work that could start; however, I had some persistent people looking to work on the bridge. One of those guys was Emery Baker, who was about 6'5" tall, 250 lb. with a bald head and tattoos all over his body and he had an earring! Now, in 2020, this is the norm, but I can tell you that in 1981, someone like this was either a hardened criminal or belonged to a dangerous biker gang. There was also something about the way Emery applied for the job—his body and facial language subtly implied that if he wasn't hired, it could mean serious personal repercussions for me! He lived locally in Avola, which wasn't that far away, so no travel time or living out allowance would be required and he was a member of the carpenter's union, so he met the basic criterion, but he scared me just a bit. Eventually, once we were ready to proceed with the concrete work, we hired Emery along with Wayne Clark, a little wiry guy who lived down by Barriere. Wayne wasn't very tall, but anyone could tell this guy was very durable and tough. He carried a hammer in his pouch that almost dragged on the ground, and the head of the hammer was twice as big as a normal Estwing. On the other side of the hammer was a primitive adze type of chopping blade instead of a standard nail puller. These guys got started on the east abutment and right away we knew they were good and enjoyed production. It became Clem's and my job to make sure they had the necessary materials to keep producing smoothly. We also hired a couple of local labourers, Ersul Fox and Jim Therres, and they both turned out to be excellent workers. Jim didn't have a lot to say but was a very steady worker. Ersul, on the other hand, was an extremely likable guy and he often chatted and told jokes to the rest of us, but he was also a diligent, reliable guy. I went to dinner at Ersul's house on more than one occasion, and he had a great home on a little acreage and a nice wife and family.

By the time the west pier caissons were installed, the east embankment and abutment were complete. Concrete work then continued on the west pier into the winter while the work bridge was still in place, and Franki completed the work bridge on the east side and started those caissons. Bob Belly was the superintendent for Franki. He was a very capable bridgeman superintendent and we got along pretty good, even though he was a Frenchman who loved to borrow my office phone and speak in French in front of me to his boss, Chuck Forcier, despite the fact they both spoke fluent English. Bob always wore an open shirt showing off his bushy, grey hairy chest in any weather, even if it was snowing and 10 degrees below zero. Doug Mace was Franki's foreman. He was a great guy who loved his job and was always smiling or laughing. Doug and I became good friends, and we went fishing a few times. On one day in particular, we caught about 20 rainbow trout while ice fishing on a nearby lake. I worked with Doug again several years later when I was working for Quadra Construction.

After the west pier concrete was finished, that work bridge was removed. Our crews worked right through that winter, finishing the east pier and finally the west abutment, heating, hoarding and shovelling snow all the way through. It was a very heavy snowfall year. One morning, I got up and could barely find my truck because it was buried above its roof! Once the piers were finally completed and the west abutment backfilled, we were ready for the concrete I-girders to be installed. They were manufactured by Gulf Concrete, which was the last project that they undertook, as far as my memory goes. They had an interesting method of installation using a specially built launching truss. The launching truss was basically a square welded steel lattice framework that was assembled on site, then cantilevered from one abutment right out to both piers. Once the truss was in position, a rolling cart was set on top of it and one end of the concrete girder was fastened securely to it. The crane located near the abutment then carried the other end of the girder and launched it out towards the pier. Once the leading edge of a girder reached the pier, that end was hoisted up from a headframe that was previously installed across the length of the pier cap and moved by trolley to its final location. While this was going on, the other end of the girder was positioned onto the abutment bridge seat by the land-based crane. This was a good procedure, albeit a bit slow and dependent on hand chain falls to hoist on the piers, so it was very laborious. This process continued until the east and centre spans were erected, then the launching truss was de-launched off the bridge, taken apart and re-installed on the west span for the final six girders. Once the final girders were installed, the headframes could be removed off the piers.

Erecting Girders – Clearwater Town Bridge

Now came the labour-intensive deck formwork, which consisted of diaphragms between the girders, interior soffits and overhangs. Clem had plenty of experience with bridge decks, and I had seen quite a few built in Campbell River when I was an inspector, so we were both keen on making this a super-efficient operation. The job was already going to make big money, mainly because of the river pier design change and some lucrative changes in quantities, such as bridge end fills and rip rap, but it was our goal to smash all of the estimated man-hour productions for all the labour items as we had already done on the piers and abutment construction. The control estimate for the bridge deck items was merely a target, and we intended to destroy those man-hour productions! The first step of the deck formwork was for the diaphragm walls that were between the girders. They were already tricky geometric shapes to match the cross-section of the I-shaped girders, but they became much more difficult because every diaphragm was skewed differently due to the curve of the bridge, so there were no two sections alike. We assigned Emery Baker, Wayne Clark and Des Newby to a team and Clem and I fine-tuned their methods from day one. Basically, we set up a race between Wayne and Emery who were working on their own separate row of diaphragms with Des working above, cutting pieces of material as Emery and Wayne called the dimensions and angles out. It worked fantastically! The guys were loving the great production and setting very high standards with the quality as they went. In the end, when all the dust settled, the boys had beaten the labour man-hours by 25%, which was huge for this type of complex, angled formwork. The final man-hour factor achieved was 0.148 MH/SF (1.59 MH/m^2). I remember Clem doing manual daily costing from the beginning of the work on this job, as he had done on the Trout Creek Bridge. Clem was always keenly interested in knowing his daily costs and man-hour productions, so he and I were on the same page when we got the monthly cost reports. Clem's interest in man-hour costing was remarkably different from most carpenter foremen I'd worked with, who had no use for these productions and believed they were already doing the work as fast as possible, so why did it matter?

Once we started forming the interior soffits and overhangs, we accomplished very similar competitiveness with the crew and we beat the budget for interior soffits by 26% and the overhang budget by 43%. Needless to say, we all enjoyed the high production and good comradery, and we would often offer to buy the beer at the pub on some of those nights. Wayne Clark proved himself to be most likely the most productive carpenter I ever worked with. Emery, when fully focused and engaged (as he was on the deck formwork) wasn't too far behind Wayne.

Once all the formwork was in place, the deck rebar was installed and we started setting haunches and final screed settings using Goodbrand's own Gomaco 450 deck finisher. As I recall, Harold Chick ran the deck machine and we had several of Goodbrand's concrete finishers in our crew as well, including Earl Hearnstead and Dale Richardson, whose nickname was Alki. For the substructure concrete, we were able to utilize the local ready-mix plant ran by Allan Bolster, although our assistance was required to achieve the Ministry quality standards. However, for the deck concrete, where the quality is much more crucial for long-term durability, we were required to produce our own aggregate locally out of a Ministry pit and we batched the concrete ourselves using a portable Fastway concrete batch plant that we set up in the Wells Gray Hotel parking lot about ½ mile away.

The final section of the three pours for the deck was completed on a beautiful, warm, sunny day on August 26, 1982, and the reason that I know that is because my second child, Taralyn, was born that day! Annette went into labour during the night, and I dropped Derek and our pet beagle over at Rocky and Shelley's house, took Annette to the hospital and then went to work. I was busy running between the concrete batch plant and the deck pour, with everything going quite well, when I got a call on my radio-telephone saying that I should head up to the Clearwater Hospital, which was only a few blocks away, to assist with the birth. After Taralyn arrived a couple of hours later, I went back to work and we finished everything up. Harold, Earl and Alki did a great job of all the finishing and the rest of the crew worked very hard to make it a huge success!

Anyone who is in bridge construction knows that when the deck concrete is completed, there is a party and everyone unwinds and has a good time. However, this was unlike any bridge deck party that I had ever seen or would ever see again! The boys were pretty stoked after three long days of pouring and everything went extremely well, but those pours are never without stress. Coupled with the birth of my new daughter, our success had everyone was in a celebratory mood. It was a Thursday night at the Wells Gray Inn, but it felt like a Friday night for the crew. I supplied plenty of cigars that night and the drinks were flowing with everyone hooting and hollering. I know Alki ended up literally hanging upside down from the chandeliers, or more accurately, the light fixtures! Luckily, there wasn't that much intensive work required the next day other than tearing apart the deck machine and maintaining the water system to cure the deck concrete.

After the sidewalks were poured it came time to strip the deck formwork. Because we were over water with limited access to the underside of the bridge deck, we used a very cool stripping scaffold that Goodbrand had devised for a previous bridge deck. This scaffold was essentially a big platform made from open web trusses and outfitted with decking that was suspended from two travelling dollies riding on the deck surface. It made stripping very easy and quick, but it involved some tricks to get the scaffold around the two river piers. After hours of thinking, planning, head scratching and calculating, Clem and I decided to strap in several empty 45-gallon barrels underneath the deck so that the whole platform would float. I might have even tried calculus to determine the number of barrels required. (If in fact, I did, it was the only time I ever used it in my life!) So, the idea was to connect the platform to four separate cables from a tow truck sitting on the Bailey Bridge just upstream from our bridge, then lower the platform down into the water from the two dollies on the deck. Once the river's current took the platform downstream so that the tow truck lines were tight, we could then disconnect entirely from the dollies above. Two of the cables running to the tow truck were directly upstream holding the nose of the platform against the current and the other two cables were fed around the pier first and then onto the nose of the platform. Once the platform was allowed to float down stream of the bridge, the two cables from around the pier were then tightened and the other two cables were gradually loosened, thereby steering the floating platform into the next span. Everything worked perfectly up until that point. The next task was to disconnect the first two cables from the nose of the platform so the other two cables could pull the platform upstream into position to reconnect to the dollies on the deck. I was operating my aluminium fishing boat from the stern with Jim Therres as my deckhand in the bow, and we manoeuvred carefully into position to disconnect the shackles on the cables when the boat caught some of the current sideways, lurched into the taught cable and quickly flipped on its side dumping both Jim and me into the swift current! As this occurred immediately upstream of the floating platform, we found ourselves tumbling downstream under this 60'-long platform and when we finally floated up, we individually grabbed onto long ropes that just happened to be tied to the downstream end of the platform and were trailing downstream. This allowed us to pull on the ropes and climb back onto the platform. Believe it or not, the boat had an oar jammed under one seat that projected out and above the gunwales and that projecting oar got caught on the front face of the pier as the boat was sideways, so we quickly hooked a line onto the boat and we were all okay. No injuries, no lost or damaged property, but if this had happened in 2020, there would have been mountains of paperwork, investigations and maybe even some jail time! I know all this happened on October 2, 1982, because when I got home that night soaking wet to my son's second birthday party, all he could say was "Dadda fall in the wawa", and I will never forget that incident or what Derek said that night. That was definitely a close call for Jim and I!

The Day We Got Wet – Clearwater Town Bridge – Note the Boat on the Pier!

The rest of the deck stripping went extremely well with high production and without further incident. Once the forms were stripped, we installed the parapet railing and deck joints, and we were ready to open the bridge.

Ribbon Cutting Ceremony – Clearwater Town Bridge
Left to Right – Earle Hearnstead, Ersul Fox, Jim Therres, Brian Sorli, Randy Paul

The only work left was the removal of the Bailey Bridge, which was part of our contract. I had sought some advice from Art Garrison on the art of working with Bailey Bridges. Art was from Victoria and was an older retired Ministry of Highways technician, who had worked on many Bailey Bridges over the years. He visited the site and was instrumental in getting us up to speed with the methodology and components required to de-launch the bridge and disassemble it. Basically, we jacked up the entire bridge and added special rollers below the bottom flanges, then added a light, angled nose section to the east end of the bridge and winched the entire bridge west with a D8 Cat. Every 20' we would stop and dismantle the components of the bridge using a small rented boom truck from Kamloops, then winch again, disassemble and on and on. This system was working great and everyone in the small crew enjoyed the work. At one particular time during the dismantling of the Bailey Bridge, I was in the field office across the river doing paperwork or on the phone doing my regular business, when the boom truck operator made a fateful decision on his own, without direction or approval from anyone. He moved the crane about 20' further west so he could dismantle two 20' sections at a time, rather than one, which he thought would be more productive. Unfortunately, he apparently forgot about the overhead powerline that he had just moved closer to, and when he reached to get the second 20' section, the boom came into contact with the power line. The resulting arc of energy ran down the crane boom and partially melted the truck's tires, but it also went down the hoisting wire rope and instantly killed Ersul Fox, who was holding the wire rope sling, ready to hook up the next component. Other people on the ground received a jolt but were not seriously injured. When the contact was made with the powerline, my office lights flashed off and on and there was an audible bang nearby. I opened the office door and could see across the river to the accident scene. When I arrived at the scene less than one minute later, it was clear that Ersul was dead and there wasn't anything that anyone could do for him. We called the ambulance and police. BC Hydro also came to the site and we were able to move the crane truck to get it away from the line. We were all in shock and deeply saddened. Once the ambulance left, the police completed their report and confirmed that they would contact Ersul's family. We shut the job down as there were ongoing investigations and nobody felt like working anyways. The whole crew was decimated. The day after the accident, Ersul's adult son came down to the site to see me and that was a very sad and emotional conversation. By this time, I felt fully responsible for Ersul's death. I could have set barriers in place so the crane stayed where we initially set it up. I could have admonished the operator ahead of time to never ever move the crane (but how was I to know that he would move the crane on his own??). I could have stayed with the operation every minute we worked. I would never have allowed the crane to move because of the clear danger with the powerline because the entire erection procedure was based on locating the crane where I did, for that very reason. In any event, I was very upset, but Ersul's son told me not to blame myself—that wouldn't help anything. He was very comforting to me, as tragic a loss that it was for both him and me.

When I reported the fatal accident to Goodbrand's safety officer, Tom Foster, he immediately came up to the site from Langley, helped me with the process and also comforted me greatly. His lasting words were, "What doesn't kill you makes you tougher." I was very thankful for Tom's guidance and compassion. As he was a former RCMP officer, I am sure he had been involved in many fatalities, so he knew what I was going through. That was such a terrible way to end what was otherwise a fantastically successful project. In the end, the fact that we completed the bridge on time and way under budget did not seem to matter at all anymore.

For the record, we made $905,000 on that job, which was over 30% net profit, which equates to $2.44 million 2020 dollars. I know that I have always liked to take most of the credit for this huge profit as Clem and I smashed just about every one of the budgets on this job, but to be fair, the big money was made in the competitive bid itself and the pier foundation design change, so Steve McAlister, Vern Dancy and even Jim Goodbrand deserve the lion's share of credit for this hugely successful project. Notwithstanding the foregoing, I had become a good superintendent over the last couple of years of running projects and had developed good work habits much like I had to do to get through the first year of college. Problems in this line of work arose almost daily, and I learned to solve or extinguish them quickly by whatever means necessary, which sometimes included working very late into the night to come to a solution. Once the solution was at hand, I most often committed to it by writing a plan. That way I could go home and get a good night's sleep and not worry about the problem. The entire idea was actually to avoid the big mistakes or "clusterfucks", which can easily develop in this industry if you are not careful and constantly on top of things. By the way, Google defines a "clusterfuck" as "a disastrously mishandled situation or undertaking", but I would add that in this business a clusterfuck can also end up costing exponentially more than originally budgeted, so they are to be avoided at any effort! I used this practice of problem recognition and problem solving successfully for many years and it turned out to be an excellent coping mechanism for a detail-oriented guy. I should also add that being a superintendent was often a very difficult, challenging and stressful job, but when a project was well managed and it was going well according to plan, it was also an extremely enjoyable, satisfying and rewarding occupation.

While I worked in Clearwater, we had many friends and guests come up and stay with us, and we were often out water-skiing on Shuswap Lake, touring Wells Gray Park, skiing at Tod Mountain (now Sun Peaks) or trout fishing in the surrounding mountains above Clearwater. On one memorable weekend, Annette's sister Louise and her husband, Glen Hanlon, the Canuck goalie, came up to stay with us for a few days. Glen was an avid fisherman, so we were up in the mountains at a nearby lake, fishing eastern brook trout. We had a great day out on the lake, and on the way home, we were talking away, telling stories while driving on the narrow old logging road. All of a sudden, Glen accidentally drove his Blazer off the road and down a steep bank, crashing into the spruce trees that eventually stopped our downward flight. Once we stopped, we were maybe 50' down the steep bank. Luckily, neither us nor Glen's dog, Brandon, were hurt. We climbed back up to the road and had to walk out of the bush and onto the highway. Luckily, I didn't live that far away. We called a tow truck to come and pull us out of the bush, and we got a real rocket scientist. As we scampered down to the truck, we all smelled gasoline, but he was puffing away on his cigarette like it was nobody's business, so we were both worried about everything catching on fire. Eventually, we rigged up a couple of snatch blocks to some trees above and were able to slowly get it up without any damage, but the engine would not turnover, so we concluded that the cylinders were filled with oil, or maybe even gas, from lying on such a steep angle. We had the truck towed to the garage, and the next morning it started just fine and actually wasn't really damaged that much in the crash.

Glen and I were quite close in the 1980's and early 1990's and hung out together a lot. He is also Taralyn's godfather, so there were lots of family occasions together, whether at their beachfront home in Point Roberts or at our house in Newton. We would often go to see the Canucks play at the Pacific Coliseum, and in those days, we could all go down to see him in the changeroom after the game. It was a special treat for the kids to meet Trevor Linden, Pavel Bure, Gino Odjick and the rest of the team in that environment. That is most likely why Taralyn is such a big Canucks fan to this day, some 30+ years later. In one instance, Larry Ashley, the Canucks' athletic trainer, worked on my back in the early 1980's in the athletic room of the Coliseum where he hung me upside down from the ceiling in ski boots to try to relieve the pain of an apparent compressed disc.

In addition to those who worked on the crew, I also met many local people and made many good friends in Clearwater, more so than in any other town that I worked in before or afterwards. Wayne Strobbe, Bill Bond, Colin Blair, Randy Carter, to name a few, were great friends, and we went camping and fishing many times together while I lived there. We have kept in touch over the years, actually travelling to Hawaii together once for my 30th birthday. My aluminium boat was always stored at the tool shack on the jobsite so it could be loaded in the back of my pickup in mere minutes and we could be trout fishing within 20 minutes of leaving the site. Azure Lake was another great boating, camping and fishing trip that we often made together (and hope to do again in the future). Unfortunately, Wayne Strobbe died of a heart attack in his sleep several years after I left Clearwater.

In the end, working again with both Clem and Rocky on the team on the Clearwater Bridge was a great pleasure, and it made the job a lot of fun. Clem and I remained great friends throughout this job, and we would often spend time together dirt biking in the mountains, woodworking in his well-appointed shop or going out for dinners together with our wives. Other than the disastrous event with the powerline and losing Ersul, the Clearwater experience was one of the best in my career on so many fronts.

Murray River Conveyor Crossing - Tumbler Ridge

After working out of town in Clearwater for 16 months, we returned to our home in Newton on December 16, 1982, now with two kids, and I reported to the head office in Aldergrove. I seem to recall that I was assisting Vern Dancy estimate a big foundation job for Canada Place in Vancouver. It was a nice change of pace, and it felt good to part of the larger Goodbrand team, getting to know everyone in head office. One strange recollection I have of that stint in the head office was having to literally "sign out" new wooden pencils from Henry Wickert, the shrewd and frugal company accountant who instituted this rule in an effort to make everyone aware of the high cost of office supplies! I found this practice to be quite bizarre, having come from the field where I was trusted with literally millions of dollars to meet or beat a budget.

I enjoyed the team atmosphere of the head office. Sometimes Jim Goodbrand would pick up some beer and a group of guys would sit in the lunchroom after work and swap stories. On one occasion, the night progressed to leg wrestling, which was a riot! It was great team building! I think I was even issued a minor amount of shares to participate in profit sharing at that time. However, I wasn't there long when senior management (Jim Goodbrand and John Sinnema) openly wondered aloud what this young pup was doing in head office: "How is he going to earn his keep not being out in the field and on a job?" I guess the philosophy wasn't too hard to understand—if you have a strong and healthy horse, make sure you have him hooked up to a heavy plough. On December 28, 12 days after returning from Clearwater and three days before New Year's Eve, I flew to Dawson Creek and took my first helicopter ride into Tumbler Ridge to start my new assignment.

The new assignment was to be superintendent of the Murray River Coal Conveyor, which was essentially a bridge over the Murray River used to carry the big coal conveyor from the Quintette Coal Mine down to the processing plant. The bridge crossing was a very critical piece of the entire mine project, and the mine owner was very concerned about us meeting the schedule, so Goodbrand scheduled two, 12-hour shifts per day, working seven days per week! This was a great job for an hourly guy with overtime pay, but I was on staff so that meant do whatever was required for the benefit of the project and the company at my comparatively meagre monthly salary. John Wigle was the project manager for Quintette Mines, and Kilborn were the engineers on this project. John, who was a very easy-going and likable guy, would often take me aside and tell me that every day we lost in the schedule, the mine would lose $225,000 (1982 dollars!), so that sort of set the criteria to plow forward at almost any expense. It also helped me get prompt approval of justified extras, as long as delays were avoided. I lived in the 1000 Man Camp, and what a life that was. I was in charge of crews that were working 24 hours a day and seven days per week, and I got to hear guys in the room next to me farting, coughing and snoring all night! The food was excellent though—it was like eating in a four-star cafeteria. One problem with being in charge of two shifts was that you sometimes got a knock on your door at 3:00 AM when something went wrong. Luckily, I had several capable staff to assist me, including my assistant superintendents, Bill Campbell and Jim White; our surveyors, Skip Singel and Earl Whittemore; and Darryl Salanski, who was an excellent project engineer. Probably the most fun I had in camp was playing basketball in the gymnasium with Drew Copley one night. The fun part was that they had 9'-high hoops (instead of the regulation 10'), so we could both act like Michael Jordan dunking everything! The next day, I could hardly move my right arm from so much dunking!

The first task at hand on the site was to install a temporary work bridge across the partially frozen river. I think the original plan was to utilize an ice bridge at the location; however, at this latitude, the use of ice bridges can be very risky, so the decision was made to install a conventional steel skid bridge with timber piling for each abutment. The suspension conveyor bridge had fairly typical concrete foundations, but the subsurface design of the conveyor crossing was anything but typical as it seemed that a few PhD-level soils engineers were involved in the project. Quadra Construction was Goodbrand's subcontractor on the job, and they were responsible for the installation of hundreds of H-Piles in those foundations. Clancy Lannon was the foreman for Quadra, and I loved working with him! He was a Newfie, and he was always smiling. It was fun solving any problem that arose with Clancy. Mike Nightingale was the project engineer for Quadra, and while he was on the younger side, he was also a very capable and likable guy to work with.

The soils in that valley were primarily silty soils without great load-bearing capacity, so all of the foundation piles were fitted with special driving shoes so they could be driven into the soft rock at considerable depth. Constant experimentation by Dr. Dirt (as the soils consultant became known) resulted in more piles being pulled to examine the driving shoes than were actually installed on some days. The staff joked that we should have a unit rate for pulling piles as well as driving them. Generally, the piles were installed using a vibro hammer for the first length, then they were spliced and finally driven with a diesel hammer to final grade seated into the rock. The bridge foundations were designed to resist ice flows as well as all other lateral forces, so many of the piles were installed with fairly steep batters, making everything much more interesting.

Quintette Mine – Murry River Conveyor Crossing

Because this project spanned over the winter months, once the piles were installed, frost protection had to be in place to prevent the foundations from freezing, and that included clean, crushed rock, double layers of blue Styrofoam and insulated blankets as well as completely enclosed and heated foundations. The suspension bridge was guyed from several large concrete blocks upstream and downstream of the bridge. These guy anchors were large, angular shapes to resist potential ice flows in the river valley, and they were very challenging to form and pour. The concrete was supplied by the Goodbrand batch plant a short distance away at the M19 pit, and all of the foundations required extensive heating and hoarding before and after placement due to the cold winter temperatures. I learned on that job that the ideal and most efficient working temperature for tradespeople in that type of construction was -5°C. At this temperature, there were no bugs and a carpenter could still handle things efficiently with his/her bare hands and the ground was just frozen enough to avoid wet or muddy conditions. However, at this temperature, the ground would not be so frozen as to affect most excavation and backfill operations, and placing concrete at this temperature was great for reducing the negative effects of the heat of hydration but not cold enough to cause a great deal of heating and hoarding costs. When temperatures got colder, as they certainly did, most efficiencies dropped off fairly quickly and hit rock bottom at about -30°C. At that point, although we had to persevere on many occasions, efficiencies could be horrible; however, the cost of not proceeding was often more than just continuing as best as we could.

Anchor Blocks – Murray River Conveyor Crossing

At the peak of this project, we had 95 men working on it, between Goodbrand, Quadra and the reinforcing steel subcontractor. To the best of my recollection, this project was originally bid at $3,500,000 and was to be completed in three months. By the time we were complete, with all the extras and design and scope changes, the revenue was almost $4,500,000 and we ended up making $1,100,000 profit over the extended four-month schedule (This was huge money in 1983!). However, when I finally returned to head office, I was informed that the company overall didn't have that great of a year, and there would be no bonuses again! The rumour was that the tailings pond earthmoving project at the Quintette mine did not do that well, but that was a very intensive company equipment project, so a lot of the costs incurred was company rent literally paid from the left pocket to the right pocket, so who knows how bad that job was??

The Completed Conveyor Crossing

I will digress at this point to give my unsolicited opinion on how Goodbrand structured its company organization to accomplish this magic profit-disappearing trick. As far as I know, all of the company-owned equipment was owned by Goodbrand Industries, a company wholly owned by Jim Goodbrand, but all of the contracts were undertaken by Goodbrand Construction, which was owned by shareholders. Jim owned 75% of those shares and key employees owned up to 25% of the remaining shares. So, the net result with this set-up was that Goodbrand Industries could set their own equipment rates for Goodbrand Construction to pay when the equipment was being used on Goodbrand jobs. The use of that equipment was not optional if there was company equipment available, so there was no competitive marketplace to allow jobs to become more cost-effective.

The other piece of the equation was that if one was to rent equipment from Finning Tractor, for example, the hourly rates were usually based on a standard work schedule of 176 hours per month; however, if you worked six days per week and 10-hour days, amounting to about 260 hours per month, you would get a credit for ownership costs on the difference of 84 hours per month (260 – 176 = 84). Goodbrand did not offer that credit in those days. That alone could be a considerable difference in equipment costs when you have a large fleet of 641 Cat scrapers, 245 Cat excavators and D9 dozers working seven days per week on a job like the Quintette Tailings Dam Project.

Other related issues included the way Goodbrand dealt with idle rent and equipment repairs. Idle rent is intended to cover ownership costs of equipment when less that 176 hours per month are utilized with any piece of equipment. This makes perfect sense, but again, there should be an accumulating credit for all excess hours before charging idle rent to the job, which was not always the case at Goodbrand. In regards to company-owned equipment repairs, running repairs and day-to-day maintenance are almost always considered a job cost, but major repairs (like engine rebuilds, for example) are typically covered by the owner (Goodbrand Industries in this example) because a significant portion of the rental rate charged is to cover the cost of those major repairs. Again, it was very often the case that the jobs would pay for major repairs as well as a general re-build at the end of the project, which was very unfair to Goodbrand Construction and its shareholders. There were, no doubt, other financial, risk and tax considerations for Jim Goodbrand, but this process certainly did not stimulate bonuses for high achievement on the project level.

This continued lack of bonuses caused me great concern, as I was starting to feel like a beaten mule! It was my opinion that hugely successful projects that required long hours, hard work and ingenuity to make money for the company should be rewarded and bonuses should go hand in hand with those successes. I certainly wasn't there just for the monthly paycheque! One has to remember that this job was no nine to five job, and it was always way out in the boondocks for me with lots of stress and long hours. As young as I was, I had been given project assignments to complete after others previously screwed them up. In fact, I believed I had been given some of the toughest jobs at Goodbrand, and I wanted to be compensated accordingly. On one occasion, after completing the Conveyor Crossing job, Jim Goodbrand actually told me to my face that he was forced to crawl on his knees in front of the bank manager at the downtown Vancouver branch of the TD Bank to extend the company's credit line because I was spending so much money on my job at the Conveyor Crossing! Well, if you have a job with that many men working 24 hours per day, seven days per week, you should expect to be spending a lot of money, and it takes cashflow. Unfortunately, I had a hard time feeling sorry for a guy who just netted a $1.1 million job profit over a four-month period on one single job, plus all of the profit built into the company equipment rent.

While I was working for Goodbrand Construction at Carnes Creek, Fort St. John, Agassiz Mountain Prison, Clearwater Bridge and Tumbler Ridge, Jim Goodbrand expanded his company dramatically into virtually every aspect of construction and development. He also built a big, beautiful new home on a spacious, white-fenced acreage in Tall Timbers, Langley and acquired a Jet Ranger helicopter and a King Air twin-engine airplane. So, while I had absolutely no hard feelings for Jim because I really liked and respected the guy, it was just a bit hard to hear there would be no bonuses again! I was getting tired of making others rich without sharing in the spoils—so much so that my boss, Steve McAlister, and I started to have serious conversations about going into business on our own.

CHAPTER 12—
Quadra Construction

Fort Nelson River Bridge

SOON AFTER FINISHING UP IN Tumbler Ridge, a bridge substructure project north of Fort Nelson came out for tender, which was of great interest to both Steve and I due to our recent experience with similar large-diameter pipe piles on the Clearwater Bridge. After the Clearwater Bridge was completed, many of the Ministry designs were now based on large-diameter extended pipe piles instead of the cofferdams, foundation piles and conventional pier foundations, so this bridge seemed right down our alley. We both agreed to leave Goodbrand and start up our own company, but we had little time to move and had to start bidding on this job fairly quickly. We approached one building-type construction company that Steve knew locally that had bonding capacity and they initially seemed interested, but in the end, they were not comfortable with this bridge project way up north. By that time, we were working closely with John Simonett and Art Forsyth (the two partners at Quadra Construction, who we just finished up with at the Conveyor Crossing) on putting together the bid for this bridge. When it looked like Steve and I were going to be without financial support or bonding for this bid, Quadra made us a proposal. We would bid the project jointly, with Quadra supplying the bonding and the cashflow. Quadra would perform the piling work and Steve and I would take care of the overall site supervision and concrete work. Furthermore, Art offered to sell Steve and I his company shares with future profits from the company, as he wanted to retire over the next while. This sounded great! So, off we went to the site visit, 60 miles north of Fort Nelson by gravel road, and we started developing a plan to build the bridge.

Over the next few weeks we brainstormed, thinking up new concepts and ideas every day on every intricate detail about how to build the bridge the most economically. We collectively came up with some great ideas in the formation of our bid. When the bids opened on July 6, 1983, I think it was John who flew his float plane over to Victoria Harbour along with myself, Steve and Art so we could attend the bid opening. We were shocked by the results! Our bid was $1,245,275 and the next bidder was $1,758,696, a whopping difference of over a half million dollars and 41%! It was a very quiet flight home that afternoon. It looked like we would be building the bridge, but it also looked like we had missed the market on this one. Over the next week or two, we retraced every step of the estimate and did not find any errors or oversights. We bid the project just the way we planned to build it and that's why we were so low. One of the main reasons our estimate was considerably low was because our plan was to build all the river piers in the winter off an ice bridge, whereas the Ministry's original plan was to utilize either work bridges or barge-mounted equipment on the river so the work could be completed in the summer months. That type of marine construction can be very expensive. Well, to make a long story short, we bid the project around an ice bridge and that was exactly how we built it.

We mobilized to the site in late September of 1983 so we could set up our concrete batch plant, perform site excavation and commence the shore pier piling before too much frost got into the ground. During that time, we prefabricated steel pile driving templates and formwork panels for all the piers so that once the piling was installed, we would be ready to perform the concrete work as quickly as possible. The formwork for the big river piers two to five were prefabricated in Karl Schlauwitz's shop on his ranch near Dawson Creek in December, well before the pier piles were even installed, so it was very important to get the piles in the exact correct position and batter so the custom-made forms would fit. The two shore piers consisted of 20" diameter pipe piles, the river piers were built from 42" diameter pipe piles and the two concrete abutments were founded on 14" diameter pipe piles. Access from one side of the bridge to the other was provided by the barge-operated ferry service that ran a mile downstream from the new bridge alignment. At the time, that was the only way to cross the river in the summer on the road to Fort Simpson and the service would be terminated once the bridge was completed.

Setting up a concrete batch plant in such a remote northern region for use through the winter was quite a challenge. We drilled an 80'-deep water well on the site with the help of Vern Zwick, a local well driller and "water witcher" who came out with the willow sticks, but I don't think the well ever produced enough water to run the plant effectively, so we were forced to use a water truck at least some of the time. To begin the set-up, we drove timber piles into the ground to construct bins for the two aggregate products, then we laid 14" diameter pipes onto the ground to heat the aggregate from below using 1 million BTU propane torches in each pipe. Next, a Fastway silo and batcher rented from Juniper Construction in Kamloops were installed along with a diesel generator, water tanks, admixture storage and dispensing systems. Once that was all in place, we installed a rented Sprung Structure (like a large circus tent) right over the entire operation with the silo sticking out above at one end of the building. We then installed propane infrared heaters throughout the interior of the building to keep everything nice and warm. We had enough room inside to store all of the mixer trucks and the loader for batching.

The Batch Plant – Ft. Nelson River Bridge

Once the cold weather arrived and the river started to freeze, we were able to start building the ice bridge right across the river on the bridge alignment. We used gas powered ice augers to drill through the ice and pump water up onto the surface to gradually increase the thickness. After we had the ice about 4'–5' thick, we were able to run lighter equipment on the river and install raw logs on the surface that would be immersed in the ice as reinforcements, but first we had to accurately lay out the pile locations so that no logs were installed where we would later auger through the full ice thickness to drive the piles. We continued pumping water up day by day until we ended up with about 10' of ice. On at least one day that I recall, the temperature at the bridge site was –50°F (there was no such thing as adding in a wind chill factor in those days!), and it was actually too cold to build ice because it would be too brittle! Those were long, cold days out on the bridge site with little to do other than make sure the pumps were running.

On one evening just before the river started to freeze, I was heading back to town, which was about one hour's drive in the snow and I came across a hitchhiker in the middle of nowhere! He was around 30 years old, and he was looking for a ride back to town to get his last load of provisions before running downstream 30 miles in his little boat to his winter trapping cabin. Well, it was certainly no problem for me as I was heading that way. I was worried that he might have died out there had I not ventured along! He, in turn, was so delighted to have me give him a ride right to the place he needed to go, he told me I could have a front quarter of moose that he had just killed for his winter food. He insisted he had more than enough food to last the winter and told me where he had hidden his stash in his boat near the ferry landing. I picked up the frozen moose the next day and took it to a local butcher to be cut up into steaks, roasts and hamburger. Several days later, looking forward to this rare delicacy, I picked up the wrapped meat and fried one of the steaks with butter in a hot cast-iron frying pan in my hotel suite. What I discovered was by far the toughest meat I had ever tasted in my life! That night, I started to learn the nuances of cooking game as opposed to beef. At that time, I was living in the owner's suite at the Coachman Hotel, which was a nice little suite just behind the front desk. I had an exterior door installed into the suite so I could come and go without going through the front desk every time, so it was a perfectly private, quiet and comfortable place—maybe the nicest accommodations in all of Fort Nelson at that time. One of the ladies who worked behind the front desk gave me some advice on cooking moose with onions, bacon, etc., and I think I eventually got the hang of it so it was at least edible.

Once the ice bridge was finally completed, we surveyed the pier locations and installed drilling templates for the 42" pipe piles. Then, using a rented rathole machine with ice lugs from Fort St. John, we drilled holes through the ice in exactly the right location and at the correct batter for each pile. Next, we installed our prefabricated structural steel driving template on top of the ice at that location. This was a steel structure about 20' high that allowed us to insert the six 60'-long pipe piles through the holes in the ice down to the river bottom while resting on the template above, holding all of the pipes exactly in the correct location, batter and plumbness so that our prefabricated form panels would eventually fit perfectly. The next operation was to drive these piles one at a time into the riverbed using a big vibratory hammer, then the template could be removed and driving could continue until the top of the piles were just above the surface of the ice bridge. The piledriving foreman was Terry Shank, and some of his crew included Doug Mace, who I worked (and ice fished) with in Clearwater and at the Conveyor Crossing. Roland Chiasson, who was a bridgeman welder, also came from the Clearwater Bridge where he and Doug worked together.

Terry Shank, me and Doug Mace Under the Big Vibro Hammer – Ft. Nelson River Bridge

The next operation was to excavate within each pile with a crane-mounted auger on our 45-ton Linkbelt crane. To avoid a blow in from the bottom of the pile due to any differential in head, the piles had to be excavated primarily in the wet, but this was very difficult when it was averaging –30°F outside. Therefore, we adapted various excavation procedures to excavate in the dry using bentonite for the first 20' or 30', then we filled the pipe with water and completed the excavation using various-style auger buckets, depending on the material encountered. Many different auger and bucket configurations were sourced from all over the country to accomplish this pipe excavation because the soil varied in size and hardness. Sometimes, it became necessary to make inspections of the soil material at the bottom of the pile in a dry hole and we were lowered by the crane down 50' to 60' to the bottom of the pipe, which was eerie to say the least! (This task was not for anyone who suffered from claustrophobia.) In some cases, a churn drill was required to blast through large cobbles found in some of the pile locations that couldn't be excavated.

Once the pile was excavated to the required tip elevation, we placed a tremie concrete seal of approximately 10'–15' in depth. After adequate curing time for the concrete, the pipe was then dewatered and several feet of crushed rock was placed into the pile where it could be proof driven with an internal mandrel-style drop hammer. This internal hammer was basically a heavy pipe filled with steel and concrete that was hoisted up with the crane and dropped onto the crushed gravel repeatedly until the advancement of the pile was determined acceptable for the design bearing capacity. From that point, the piles were filled with mass concrete to within 8' of the ice surface. Next, the second 60' length of the centre pile was spliced onto it by welders and a new template erected on top of that pile, which set the correct batter for all of the remaining adjacent piles that were then spliced onto the driven lengths just above ice level. Once all the pile splices were completed, the diaphragm wall forms could be erected, rebar columns installed into the piles and diaphragm wall rebar installed. As this was all taking place in extremely cold weather, the entire pier was then enclosed in hoarding panels and propane heaters were installed to get all of the piles and rebar well above freezing. Clem Buettner, who again joined me as carpenter foreman, was instrumental at this stage of the project as he had some really effective ideas for hoarding the piers. His method was to build a 2" x 6" lightly braced wood frame, then place a woven polyethylene fabric on each side of the frame to create an insulating air space in between. The frames were generally built in rectangles of 12' x 24', then stacked and spliced together so the entire pier, including a roof was fully enclosed. After the diaphragm walls and piles were fully concreted, we formed and poured the pier caps, which were very similar to those at the Clearwater Bridge. The formwork and concrete went extremely well under Clem, along with Karl Schlauwitz, who I previously worked with at the Conveyor Crossing in Tumbler Ridge.

Building the River Piers off the Ice Bridge – Ft. Nelson

Having read all of the daily reports and my personal notebooks to spark my memory for writing this part of the book, I have to say that the Fort Nelson River Bridge was by far the toughest job I have ever run as a superintendent! Working in the cold at those temperatures in a breezy river valley that far north and that far out of a small town like Fort Nelson presented huge challenges. We had continual equipment breakdowns, failures and dead batteries along with frozen pumps, tracks, augers, mixer trucks, etc. We also had collapsed boom tips, damaged crane booms, damaged drop hammers and a cracked outrigger beam. It went on and on, and it's a good thing we had a local, full-time welder/mechanic named Peter Williscroft. He was a big, jovial guy who took everything in stride, no matter what time of the day or night or what temperature it was. Peter got it fixed—many nights beneath the colourful late-night northern lights. The constant repairs also taxed me greatly as I was constantly on the phone at night looking for replacement equipment, parts and supplies from all over the north and as far away as Edmonton and Calgary, and everything was required in a rush.

On one Saturday night, after a long week out on the job, Clem and I got together for a rare social evening in the Fort Nelson Hotel with the owner of Blue Canyon Concrete, the local ready-mix operator, from whom we rented the mixer trucks. As it was usually not a good idea to socialize too often with the bridgemen crew after work, we selected the small lounge where we would not likely run into them. This was a special night as I remember Clem wearing a nice brown leather jacket, and we were having a nice time enjoying a couple of drinks before dinner. Well, after a while, our batch plant loader operator came over and sat with us. He was a union-dispatched Local 115 operator, and we did not know him prior to this assignment. He seemed to be okay on the job, so we initially didn't mind him joining us for a while. As time went on, it became obvious that the guy had been drinking quite a bit and was becoming a little erratic and mouthy. He seemed to think that he knew more than Clem and I put together about the concrete plant and the bridge, and pretty soon he started pointing his finger at me, telling me what he thought. Not long after that, he was poking me in my arm and chest as he spoke, abusing both Clem and me. Finally, after an hour of putting up with this guy, he touched my face with his finger and I lost it! I immediately stood up, grabbed him with both hands around his collar, lifted him up and moved very quickly to the timber post adjacent to our table where I slammed his head into the post! As he was slumping into a heap, I was reaching back to give him the biggest haymaker in history when Clem tackled me to the ground. I was so mad that Clem was afraid I might kill him. It had been building and building for well over an hour. This guy was not only ruining our night, his drunken blathering and poking at me infuriated me. God knows I had been dealing with enough stressful situations on that job. I really didn't need this.

By the middle of April 1984, we had completed our work and all of the piers looked great with a nice job of concrete finishing that contrasted with the black steel piling. We had fully demobilized from the site as the ice bridge was melting quickly, and I parted company with all of the staff and hourly workers. I then made the long drive back to Vancouver, where I was surprised to see cherry trees blooming all over the place when it had still been winter in Fort Nelson with snow on the ground! It was shocking to see the stark difference between Vancouver and Fort Nelson, and I was starting to worry that I would be stuck in that northern construction world of winter fuel, hoarding, propane heating, winter batch plants, ice lugs on the track equipment and a constantly frozen mustache!

Being back in Quadra's head office was really nice because it was situated on a floating barge in Coal Harbour in downtown Vancouver and had yachts passing by and floatplanes buzzing overhead. The view from the office was second to none, and two of the owner's float planes were tied up to the barge if you ever wanted to get somewhere quickly.

Ft. Nelson River Bridge

At that time, Quadra did not have a computerized job costing system, so I never really had accurate knowledge of actual overall costs incurred versus the estimated costs for the bridge. As far as I know, the job did as well as we originally expected and we made our anticipated margin, but was that ever tough sluggin'!

By this time, both Steve McAlister and I were no longer interested in buying the shares from Art Forster for a number of reasons. One reason was that Quadra was tied to its union agreements with the Piledriver's Union, compared to our preferred experience with other more conventional trade unions. Steve eventually returned to work for Goodbrand, and we parted as good friends. I still had a lot to achieve in the construction world and one way or another I wanted to start my own company. Actually, Art Forster helped me set up my own operating company, which I called Delta-Star Contracting[11]. From that point forward, I was contracting my services out to Quadra under that company name. John Simonett then asked me to assist with a marine-based project that they were in the middle of in False Creek, Vancouver. The project was installing literally hundreds of foundation piles for the Preview Center for Expo '86, and it wasn't going very well. The entire foundation piling design seemed seriously flawed because the piles were very thin walled and not strong enough to be top driven like conventional piling. Hence, the piles could only be driven with a mandrel within the pile driving on a layer of crushed rock placed into the pile above the welded baseplate, but again, that connection to the baseplate was still subject to failure as well due to the very thin wall of the piling. Besides these fundamental driving problems, the sheer number of piles and relative batters of each pile, along with an extremely aggressive schedule requiring several outfitted barge-mounted driving rigs, made the execution of the project a logistical nightmare. While I enjoyed working with John and Doug Mace again, I felt that I just couldn't really help with the logistics, and I thought they were doing everything they could do to get the job done. Sadly, I gave John and Art my decision and decided to move on. I know things would have been completely different had I started this project, but it really felt like I was again being tasked with cleaning up someone else's difficulties; however, it has always bothered me that I left John and Art in that difficult position.

I built an office in the basement of our Newton home and planned to work from there for the time being.

---

11 Delta-Star was a term that I read about in an old Joseph Wambaugh novel called The Delta Star that referred to an imagined excited state of creativity called Delta to Delta-star. This name had zero to do with me coming from Delta, and I certainly never thought of myself as any kind of star!

CHAPTER 13—
Delta-Star Contracting

QUADRA HAD JUST STARTED THE Goat River Bridge project near Creston, B.C., which it had been awarded by the Ministry of Transportation earlier that spring. It involved some steel foundation piling that was nearing completion, and the concrete work was just getting started. Terry Shank, who was my original foreman in Fort Nelson and who had a background as a bridgeman, was running the job for Quadra, and I remember Art and John requested that I go make a site visit and see what I thought about how things were going. Well, I visited the site but immediately found myself in a very difficult spot because, other than the installation of the piling, I didn't like the way things were being done at all. As I recall, one abutment and one pier were partly formed with an old set of EFCO steel handset forms, which, in my opinion, was a terrible choice of forming methods. It was labour-intensive hand work to both form and strip, and it was going to leave the concrete with all kinds of form marks and nasty staining because of the age and condition of the small form panels, which would necessitate a lot of additional finishing costs to repair. My call to John was blunt: "You have the wrong man building this bridge!" Apparently, they understood, believed and trusted me because my very next phone call was to Clem Buettner, who was sitting at home, not presently working.

I cannot recall what happened to Terry, but I am guessing they put him on to another piling job they had somewhere. I am sure he didn't think too highly of me though. All I know is that with Clem on site, we chose carpenters, labourers and cement masons to undertake the work, which was far more typical in the industry than using bridgemen. The steel handsets were removed from everywhere except footings that would be buried and hidden, and the carpenters quickly prefabbed typical plywood forms for the piers and the abutment walls. That was Clem's specialty! I returned home at that point as I could assist Clem from there, if and whenever necessary. As these structures were very simple, the substructure was completed very quickly and with excellent quality. The next step for me was to assist with the planning and logistics for the delivery and erection of the pre-stressed concrete girders (which was also completed without a hitch) utilizing local cranes working from the essentially dry riverbed. Next, Clem was away to the races on the deck formwork, which was very simple compared to the Clearwater deck. The deck concrete was again finished using Fred Dickhut and his deck finisher and everything turned out very well. I was happy to assist on that project, mainly by getting Clem involved, and I am confident that the project ended up as well as it possibly could have with our changes and input.

While I was engaged with Quadra on the Expo Preview Center and the Goat River Bridge, I was also in discussions with a friend of mine from Delta, Bob Bruce, about operating a construction company together. I understood that Bob was quite financially well-off, having been involved in mortgages and financing for quite a few years. Bob was very keen to work together, with him only providing financial backing. To get started, we set up a bank account and began the process of applying for bonding and insurance, which was required on all government work. In the meantime, he had some space in the back of his retail stereo store on Scott Road in Delta where I could set up my office. The process with the bonding companies proved difficult because it turned out that most of Bob's money was not in liquid or capital assets as required by bonding companies; rather, his money was in bundles with elastic bands and stored in apple and shoe boxes. When it came time to make the initial bank deposit, Bob pulled $10,000 in cash out of an eight-track case from the cupboard and gave it to me to deposit! Now, I should add at this point that I knew very little about Bob's business background nor did I have a clear picture of the source of his income, but I became aware very quickly that he would not be a conventional financial partner. Eventually, we were able to secure some bonding and start bidding projects around the province.

On June 22, 1984, Delta-Star bid its first project, a bridge on 16th Avenue in Langley, coming in 12th place out of 14 bidders. Apparently, I was still just testing the waters. Then, on July 27, we were low bid on the Fulton River Bridge located north of Smithers. This was a bridge deck overlay job and our bid was $28,170 which was $285.00 lower than Kingston Construction, who was a very reputable company in that business. We completed that project in the first two weeks in September and everything went fine. Clem came up to help on the job, along with my dad and Bernie Buettner, who was Roy Buettner's son. Oddly enough, my dad was visiting from his home in Ontario at that time, and he decided to come for a drive with me to Smithers to see what we were doing. He ended up working along with the rest of the guys as best as he could for a couple of weeks. The toughest part of that job was placing our own deck rebar—bending over all day tying rebar was a killer for most of us! The best part of that trip was when Dad and I drove down the old Coquihalla Canyon Road between Merritt and Hope on the way home, seeing all the old KVR rail grade, tunnels and bridges, all while dodging construction spreads as sections of the southern Coquihalla Highway were just getting underway at that time.

From July 30 through to December 19, 1984, I bid 10 more jobs for a total value of $5,547,902, with individual bids up to $1,567,000. We were second bidder four times and third bidder three times, but we never were low bid again. If we were low bid on any one of those projects, our lives could have turned out different, maybe better or worse, but that's contracting for you. Actually, the final bid that I prepared to submit was on December 19, 1984, for the Cliveden Overpass in Richmond, which was an overpass over the new Highway 91 created with the completion of the Alex Fraser Bridge. For some reason, our bond was not issued for that bid, I assume due to failing to meet the bonding company's financial requirements, so that was the beginning of the end of the business relationship between Bob Bruce and myself; however, one of the cool things about working with Bob was that he was an accomplished pilot, and he owned a nice twin-engine plane. It just so happened that I loved flying myself, and I had rented small, piloted planes many times for excursions with my kids out of Boundary Bay Airport. Many times, Bob and I would fly around the province going to site visits because he was always looking for excuses to get in some flying time. On one of those trips, we went to the northern tip of Vancouver Island to look at a site and on the return flight, once we took off from Port Hardy, I flew all the way home! Bob put his head back on the backrest, closed his eyes and told me to let him know when we got close to Horseshoe Bay. That was fun because I got to fly over a lot of the old bridge sites I had worked on while I was in the Campbell River area.

On yet another trip back from Victoria, after getting quite a bit of flying time in, I wanted to land the plane at Boundary Bay Airport. Annette was in the back of the plane and Bob was calmly coaching me all the way down, but as we got lower and lower, I had trouble keeping things lined up with the runway due to crosswinds. After continually correcting, with only two or three seconds to land, I told Bob to take over, and he swung hard upwind and came down hard onto the runway! That was really nerve-racking, but we were all safe.

In early January 1985, Kris Thorleifson contacted me to see if I was interested in running a job for a contractor who was low bid for a contract with the federal Ministry of Transport to regrade all the dykes and install rip rap material around the Kamloops Airport. Kris bid the job for Canarctic Ventures, but he had no interest in running it as I believe he was already working for Goodbrand Construction on the Coquihalla Highway in Hope, B.C. Well, I wasn't that busy and I wasn't getting rich bidding jobs and being second bidder, so why not? I went down to the company's office in Delta and met Bill Katerinchuk, who was the president and an ex-Peter-Kiewit guy. I negotiated a healthy daily rate and living allowance with him under Delta-Star Contracting, picked up the drawings and specifications and was off to Kamloops in their company truck.

That job also took place in the dead of winter and basically involved excavating the face of a long stretch of the dyke along Kamloops Lake to a revised design alignment and slope, then placing a rock filter material and rip rap. Canarctic Ventures had made some preliminary arrangements in an old quarry on Ord Road just beside the airport to produce the filter material and the rip rap. Before we could get started, I had to produce, along with the help of the blaster, a comprehensive application to the Ministry of Energy, Mines and Resources for the pit development and blasting plans as well as the use and storage of explosives. I still have a copy of the five-page hand-printed submission, dated January 31, 1985, which was eventually accepted.

Canarctic mobilized a big Drilltech rock drill, a 988 Cat loader, a Cat 966 loader and a Drott 120 excavator, along with Hank Schellenburg, who would oversee the pit operations, run the loader and sort the rock. The Drilltech D25 was, in my opinion, way oversized for the job, but my role was only to manage the work; it was their decision to utilize their own equipment. The Drilltech D25 drilled 6" diameter holes, and we used mostly ANFO for the explosive, which is essentially fertilizer and fuel oil and not the most delicate blasting method for producing quality rip rap; however, we managed to produce the required quantities of rock material even though we also created a lot of waste with all the over-blasted fines and small rock.

The design engineer for this project, Duncan Hay & Associates, was most likely related to Dr. Dirt from the Conveyor Crossing project, as they created a serious science project out of that job. I had produced and placed a lot of excellent quality rip rap for the Lillooet Bridge, the McPhee Bridge, the Clearwater Town Bridge and for the job in Tumbler Ridge, where we had much greater velocities and erosion potential and we never saw so much science going on. Kerrin Spurr was the site inspector, and he was also a real stickler. He carried a tape with him and if he measured any rock greater than 8" for the filter material, he spray-painted it and made us go back and break the rock in half so it met the gradation specification of 8" or less. For the rip rap, if any one dimension was greater than the nominal size, he made us pluck it out and break it in two or discard it.

In my past experience placing rip rap on river banks, one installed it in lifts depending on the size and reach of the excavator you could use to place it. In this case, Bill Katerinchuk thought that the biggest machine possible would be the best, so they rented a Koehring 1066 from Pacific Blasting. That was a monstrous machine that took several truck loads to get it to site, then required a crane and a crew to assemble the behemoth. It was an old, tired machine, and we had nothing but problems with it in the winter environment, so we kept Jack Walch Equipment Service from Kamloops busy that winter. In any event, we got the dyke completed, and we were all finished up by the end of March, 1985. I seem to recall that the

Koehring 1066 self-ignited into flames while demobilizing, and it looked like financial combustion to me! I got paid for all of my consulting invoices, returned their truck and never heard of Canarctic Ventures again.

I enjoyed living in Kamloops, even though it was through the winter time. I met a lot of people through that job and many remained friends and valuable contacts for many years. One part of my job I still enjoyed was running a variety of equipment, as I had since I got into construction in the 1970's. On that particular job, I had a fair amount of spare time, so I often ran the 966 Cat loader, the D6 Cat and now and again I would load rip rap into the trucks using the 988 Cat loader. The 988 Cat was a huge loader compared to what I had been operating in the past, and I enjoyed it immensely. By now, I was becoming fairly familiar with running all of this equipment.

CHAPTER 14—
Goodbrand Construction

Alexander Bridge - Hope, BC

AS THE KAMLOOPS AIRPORT PROJECT was wrapping up, I received a call from Steve McAlister, who had gone back to Goodbrand Construction shortly after the Fort Nelson River Bridge, and he told me they were low bid on a big bridge job near Hope, B.C. that had my name written all over it. Goodbrand had been awarded some huge projects in the Hope area for road construction on the Coquihalla Highway, including interchanges in Flood and the Hope Princeton Highway, as well as a number of bridges and overpasses. When I was returning home from Kamloops, I believe I met Steve at the Alexander Bridge site and his excitement was contagious as it was a very interesting and challenging project. I was all in for the challenge and also looking forward to working with all my friends I had made at Goodbrand. The site wasn't too far out of town, which was also a nice change.

The new six-lane bridge was to be built alongside the existing two-lane highway at a sharp curve over an old box culvert known as Suicide Corner, about 6 km east of Hope, B.C. The new alignment, which was part of the freeway upgrade from the #1 Freeway to the Hope Princeton Highway, as well as a connection to the new Coquihalla Highway, would run straight above the canyon that was formed by Alexander Creek where it enters into the Coquihalla River. The highway through this entire area was under construction by Goodbrand and this bridge was at the base of two large glacial till and rock cuts where they were drilling, blasting and excavating on double shifts. The canyon below the bridge was formed by hundreds of years of loose soil sliding down from the mountains, being eroded by the creek above or being dumped over the banks from highway maintenance operations over a 40-year timespan, making the slopes and the base of the canyon unconsolidated and therefore, very difficult to create solid foundations in for the bridge. As such, the foundation design for the piers and west abutment was very unique as it required the mining of large diameter shafts vertically through this incompetent, sloping material down into solid bedrock. This innovative bridge design was produced by Andrew Rushforth at Graeme and Murray Consultants for the Ministry of Transportation. These mined caissons were lined with corrugated steel plates that were installed in small segments and bolted together to form the perimeter of the shaft as the excavation proceeded down within the caisson. Once the liner plates were installed, pressure grout was then pumped around the perimeter to solidify the native soils around the shaft. (More detail to follow on this method.) Once the shafts were complete down to competent rock, hollow reinforced concrete columns were placed within the shafts and then conventional structures were built above ground. Goodbrand subcontracted the installation of these four caissons to Devpro Mining, which had previous experience with this type of mining work. Bill Jackson was the superintendent for Devpro and his partners in this venture were Milos Filgas and Jerry Kratochuil.

I started working in the Goodbrand head office around the beginning of April 1985, getting set up to build the bridge and then started mobilization to the site by mid-April. The bridge was supposedly being managed out of Goodbrand's Hope field office under the direction of Jerry Nauss, the construction manager for all the highway work in and around Hope, and Ed Beynon, the project engineer. This situation didn't sit all that well with me, primarily because Steve McAlister hired me back to Goodbrand for this project, and I had worked independently for him in the past. Also, it was my style to take direct and personal responsibility for a job I was assigned to and not be a part of an overall project team, especially when most of their attention and background was in mass earthworks, rock excavation and road building with the surrounding contracts. The first rub with the Hope office came when I told them that I was going to hire an assistant superintendent who could also take care of the survey on this job for me, as it was too big for me to do on my own. Their response was that I didn't need an assistant because they had ample staff in that office to assist me. Well, having been around that office a bit since getting there, I was not convinced they had the right people with the dedication available for this job alone, so I went ahead and put a call into BCIT for a Civil and Structural Engineering technologist. Some people called me a prima donna, and others said I was not a team player, but none of that bothered me. Give me an assignment and I will get it done. I didn't need a committee to build this job. The other rub was that they were attempting to dispatch equipment to me on a day-to-day basis when I needed it. In fact, Jerry Nauss himself was involved at times, once telling me that I could not get an excavator even though I had scheduled it earlier with him. This was difficult for me because all of my plans and labour utilization changed at the whim of changes or breakdowns on other projects. Normally, if I was working on my own, there would be competitive parties wanting to get their equipment on my site, so I wouldn't have had these issues.

Well, in regards to finding my assistant, I was given a few names and resumés from the recent BCIT graduating class, and I made arrangements to meet one applicant named Noel Mankey, who appeared to have the right experience for this job. We arranged to meet at the Hope site office a few days later, in mid-May, and when the time came, a young guy came to my office and introduced himself as Art Penner. When I asked where Noel Mankey was, Art replied that Noel had just taken another job the day before but told him to come to the planned meeting and apply for the job instead. Art gave me his resumé, which included some experience with Hansa Construction, whom I'd worked with at Taghum Bridge, so I thought this young pup might be okay. I pushed Art just a bit in that meeting, clarifying that I needed someone to start immediately and that I required a lot of dedication and long hours of hard work. In other words, it was not going to be a picnic, but it could be great experience for a young guy like him. His response was that he needed the work experience badly, and he would work for free, just to get that experience. I immediately accepted his offer and told him to start on Monday morning at 8:00 AM. I liked the sound of a guy who wanted to work so bad that he would work for free! I think Art wondered if he was actually going to be paid until his first paycheque arrived some two weeks later.

Young Art Penner – Alexander Creek

As soon as I mobilized to the site, I met with Harry Sandwith, still chomping on that big stogie, who the Ministry had assigned to be the project supervisor. It was going to be great working with Harry again, as it had been about seven years since I worked for him in Campbell River. Just like I worked with Rocky in Clearwater, I knew that working with Harry would be fantastic because he was such a practical guy, and I think we had deep respect for each other.

Harry Sandwith – Alexander Creek Bridge

Around that time, Devpro was starting to assemble and mobilize its equipment to the site, and Armtec tunnel liner plates were being delivered so that we could get started on the west abutment caisson, which was the easiest access to prepare. Vern Dancy, from head office, was a big help to me, finalizing a lot of the long-term purchasing requirements, including the Armtec tunnel liner plates, which, for some reason, was Goodbrand's responsibility. In the meantime, Art and I, along with Harry's assistant, Rod Mochizuki, began to get the site survey underway for the bridge elements and started to produce a complete topography of the bridge site that would be used for excavation payment quantities. When neither Rod nor I were available, Art had to get quite inventive—by inserting the levelling rod vertically into a tall highway cone, he was able to continue reasonably accurate topography survey without assistance. (This was before the days of total stations or robots, and showed Art's initiative!) Art's first real layout assignment was the caisson at the west abutment where Devpro had set up the first Armtec tunnel liner plate rings and a jacking shield to get started with their shaft excavation. I remember Harry wanted to personally check the exact starting location of the shaft with me before he gave the go-ahead for Devpro to get started. We found that the caisson was about 75 mm out of place, which was essentially in spec for that liner, but Harry's comment was, "Well, that is a hell of a way to get started, right at the edge of the acceptable tolerance." I read Harry quickly and agreed. I told Art about the issue and asked him to remove everything, dig it up and re-survey so it was starting out exactly in the right position. While that was a delay of one day on a tightly scheduled project, I think it was a very wise move by Harry and a very good lesson for Art about accuracy and getting started in the right place. One thing I noticed about Art by this time was that he was very quick to get things done; however, accuracy is often much more important than speed, so it became a constant mantra of mine that he double-check, then triple-check his work before saying he was finished a task.

Caisson at Alexander Creek

It was around this time that I had a half serious, half comical discussion with Art to tell him that the direction he was headed to become a superintendent in this business was a very tough path, with lots of headaches, problems, responsibility, out of town work, long hours, etc., and that he should think long and hard about the direction he had chosen. Of course, Art was very keen to continue at this age and stage of his life, so the warning went essentially unheeded, but this was the benchmark that I would remind him of for the next 35 years or so, and it always ends up with a good chuckle.

Around this same time, we sold our house in Newton and bought a nice home on Orchard Drive in the east end of Abbotsford. It was a great area and much closer to Hope, as I would be commuting for the next year. Derek was five years old at that point and Taralyn was three, and I thought the area would be very a good place for them to go to school.

Back on the bridge, on May 29, I requested Clem Buettner from the union hall to be my carpenter foreman, and he was dispatched immediately. We had a lot of planning to do together to get the formwork designs completed for the substructure piers and abutments. It was going to be a lot of fun getting Clem and Harry back together, and it turned out that Clem liked Art as well. Art became a regular target of Clem's (actually the Buettner family's) famous sarcasm and pranks. One particular evening after work, we all went to the stripper's show at the Hope Hotel, and I think Clem somehow dared/enticed Art into a Caesar-drinking contest. As I was driving home to Abbotsford that night with Art as a passenger, Art's head looked like a bobblehead. I think Clem easily won the contest.

While Devpro got underway on the west abutment around June 1, we started building access pads and a road down to the west pier with excavators and a dozer. The terrain was pretty steep, and we had lots of material on the slopes that we could use for access roads but we had to essentially build the roads in fill rather than cut so that we did not destabilize any of the existing slopes. We ended up with an access road down to the west pier at a grade of about 10%; however, the access road to the east pier could not be effectively built at less than about 23%. There was lots of discussion by the Hope office, head office, etc. about my plans for the construction of that access road and the crane pads—to the point that I eventually brought in Frank Main, who was a well-respected, retired earthworks expert from Peter Kiewit, to look at what I proposed. To complicate matters in that area even further, the east abutment was so close to the existing highway that we required a pretty substantial vertical shoring wall to get down to the footing level. That shoring wall was constructed out of logs and wire rope and had buried logs as deadmen, similar to what we did at Carnes Creek. In the end, the east pier road access was so steep that we eventually decided to pave it with asphalt to increase traction for the vehicles and improve maintenance. We were able to do that pretty cheaply while Goodbrand was paving some local diversion roads in the area. Art and I worked closely together on the design of the access roads, with Art performing most of the survey and producing hand-scaled drawings and cross-sections (before AutoCAD). The end result showed that the access roads would be more than adequate to get the cranes, concrete mixer trucks and highboy trailers into position as required, and it all worked out just as we planned.

Late one afternoon, around this time of year, I needed to speak to Harry about something that we were planning to do right away, but I wanted to clear it with him first. I had just seen his truck leave the site but this was before cell phones, so I had no way of getting hold of him. Art and I were carpooling at that time, so we quickly hopped into my Chevy S10 work truck and started chasing after Harry down towards Hope. We didn't see his truck at his office, so we thought he was heading to his home in Abbotsford, so we pursued further, increasing speed a bit to close in on him. We were in the Flood area where Goodbrand was just in the middle of many road changes and as we were flying onto an on-ramp to get on the freeway, we were all of a sudden airborne! The truck left the road at a high speed as I was looking over my left shoulder to merge onto the freeway. Apparently, the lane I was in was no longer a merging ramp, but a 90 degree turn with a stop sign at the end of it. We landed in the weeds with a major thud and came to an immediate stop! We did not roll the truck or hit anything, but the headroom in the cab was somehow reduced as we had bent the frame of the truck on the rough landing. Not wanting to sit there in the public's view any longer than necessary, I radioed my good friend Al Beamer, the maintenance superintendent at the shop. Without transmitting to everybody that I had been in an accident, I calmly requested he send out a loader and a chain to this location, which he promptly did. Unfortunately, the truck was a write-off, but luckily, Art and I were perfectly fine.

As was company policy, I had to fill out an extensive accident report and hand it in to the Hope office, where it would be forwarded to Tom Foster,

Goodbrand's safety and security manager. In order to submit an insurance claim to ICBC, I also had to report the accident to the local RCMP detachment in Hope, so I provided a copy of the Goodbrand Accident Report to them as well. Upon review of that report, the police officer thought that he should give me a ticket for driving with undue care and attention, which came with six demerit points. I was able to convince the officer that driving too fast for road conditions might be a more suitable ticket as that was only three demerit points. After a week or so of thinking about it, I decided to take the ticket to court and contest it. Perhaps the officer wouldn't show up or the judge might be more lenient. Well, when I went to court in Hope that day, and explained the situation to the judge, she immediately threw out the charge based on the common legal principle that you cannot provide evidence to convict yourself. In other words, the police had no evidence of their own that I was driving too fast for road conditions because they had not produced or even corroborated any evidence supporting their charge against me. I will add though, that I drove very carefully through the Hope area until we finished that job for fear of running across that officer again!

Meanwhile, back on the job, the method Devpro devised to install the caisson liner plates was to start with a ¾"-thick steel plate ring about 8' high and about 12' in diameter (referred to as a jacking shield), which would be installed just outside of the ring of the Armtec tunnel liner plates. That jacking shield would be jacked down off the installed liner plates above in order to advance downward with the excavation and ahead of additional liner plates being installed. To get started, the initial 10' of liner plate was preassembled perfectly round above ground, then installed into an open excavation with the jacking shield installed just outside of that. The perimeter excavation outside of the jacking shield and liner plates was then backfilled and thoroughly compacted and a concrete starter slab was placed at the top of the liner plates to secure that structure in place. The excavation then proceeded inside of the caisson. Once a round of excavation was completed, they jacked the shield down to the bottom of the excavation and additional liner plate segments were installed until the full circumference was in place. Excavation was mainly performed by a digging attachment from a Bobcat that was attached to two separate brackets on the jacking shield so the digger could be moved from one bracket to another to make the excavation easier. All excavated material was hoisted out of the caisson in buckets by our crane, which was located above. As the depth of new liner plates was installed, the design required that the annular space between the liner plate and the outside soil be pressure-grouted regularly through grout nipples in the liner plate to ensure that no erosion or unravelling of the surrounding soil occurred. The west abutment caisson, (the first one installed) gave us lots of problems because it was difficult to keep the shaft perfectly plumb and circular, and as the excavation proceeded downward, it wanted to wander off-line. There was lots of hand-wringing and memo writing, and many site meetings were held to try and resolve these problems, but they tended to continue throughout the job, despite various improvements to the methods. Once we got to the pier caissons, there was even more tendency for the entire caisson to slip down the slope in the poorly consolidated material, which again, caused a lot of concern.

Once the west access road was completed in mid-June, we walked a Bucyrus-Erie 60-ton crawler crane, that we had rented from Forsyth Equipment, down to the west pier with Leon Albrecht running the crane and a D8 tied to it from the rear for safety. Then Devpro got started on the one caisson there while our crews got going on the one pier foundation on rock, and both were serviced by that one crane. That caisson was designed to be 20 m deep but ended up being almost 25 m, with the last 6–7 m requiring chipping and blasting until competent bearing rock was encountered. Once Devpro had completed its work on that caisson, we took over with the installation of rebar and installed an EFCO round steel void form in the caisson so we could start placing the annular concrete column inside the shaft. That form worked very well and the cycles went pretty quickly. Above ground, we got underway using standard EFCO plate girder panels to form the columns, tie beams and pier caps. The pier cap support system was composed of two 36"-deep, wide flange steel beams supported by four 50-ton hydraulic jacks mounted to each column. We rented the beams from Brittain Steel, our subcontractor responsible for fabrication and erection of the steel superstructure. Those beams were custom reinforced on the top flange to meet our load requirements and then they were used again by Brittain in their own erection scheme, so this was quite economical for everyone.

Clem on the Right with Leon Albrecht, the Crane Operator - Alexander Creek

By October of that year, Clem and his crew had completed the west abutment and the west pier and they were well along with the east abutment. We were also well underway on the east pier as Devpro was off-site by his time. We had rented a 125-ton American crane with 120' of boom and a 40'-long jib from Canron to service our east pier from behind the east abutment.

West Pier Formwork – Alexander Creek

Devpro's work installing the caissons took longer than originally planned, partially because of problems with equipment breakdowns and the chosen methods of installation, but also because some of the shafts had to go much deeper to locate competent rock. Unfortunately, we were unable to get started in any location until Devpro's work was completed, and they were demobilized from that area. This definitely impacted our ability to complete the contract on time, and the Ministry was not in a position to allow any delays due to the political nature of the Coquihalla Highway and its timed completion for the opening of Expo 86.

On November 1 of that year, we had a serious rainstorm and the west access road looked like it was going to wash out. The reason I know it was November 1 is because on the day before, October 31, 1985, I had a back operation called a transcutaneous discectomy. They performed an orthoscopic procedure on the herniated disc in my lower back, literally grinding out the centre of the bulged disc and then vacuuming it out, enabling the disc to shrink away from my sciatic nerve. After being in constant nagging pain for the last year, I felt so good after staying overnight in Shaughnessy Hospital, that I went to work the very next day without any pain at all. Then the rainstorm hit. While gingerly walking down the west access road, I grabbed a shovel from someone and pretended that I was going to start digging a drainage path for the rushing water, when Clem grabbed me, took the shovel away and physically escorted me to the office. Luckily, the crew quickly got the idea and everyone pitched in manually with shovels because it was so wet and too dangerous to put a hoe on that slope. Ultimately, their work saved the entire road and access pad.

The Big Rain – November 1, 1985

On one Friday night after work around this time, Clem, Art and I decided to go have a couple of drinks with Rick Ross and his ironworkers at the Hope Hotel. Well, we had a hoot, and of course Art, being the young guy in the crew, was again the target of competition when he made a bet on something against Clem and lost. Unfortunately, losing that particular bet required him to get up on the stripper's stage and lick the pole! Not to lose total face with the crew of ironworkers, he complied, to the howls of everyone!

Just as the first snow started in early November, Brittain mobilized to the site with their two 165-ton Lima truck cranes and planned to start erecting girders on the west end of the bridge while we finished the east pier and east abutment. Gary Mitchell was Brittain's project manager and Pat Charette was the site superintendent. I worked with Gary mainly on the initial delivery and erection schedules, and I was more involved with Pat on the site erection. He was a very likable, competent and highly experienced structural ironworker. He was an easy-going and collaborative guy, making it easy to work with him. We had already experienced some delays with Brittain in fabrication and their original completion date for the erection (December 31, 1985) was now in serious jeopardy, again adding additional pressure on us to make our contract completion date.

Brittain Steel Erecting the First Girder – Alexander Creek

Other than normal fabrication delays, which are certainly not uncommon, Brittain also had some delays in getting the girders to the site. Their original plan was to load the girders onto barges on the Fraser River directly from their shop located in New Westminster, tow them to Hope and then transport them by trucks and steering dollies over that short distance to the bridge site. However, 1985 was apparently the driest summer and fall on record and there was insufficient draught in the river to get the barges into Hope. Ultimately, with some serious help from the Ministry, they were able to truck them to site, but they had to leave off the nelson studs on the top flange to reduce the overall height. That was cutting it close!

Once they were all rigged up on site, Brittain erected two of the temporary beams that we had previously rented from them for our pier cap formwork, and they installed those between the west pier cap and an existing old concrete retaining wall running along the highway. Once those beams were in place, they installed a rolling trolley system on those beams so the east end of the west girders could be placed on the trolley while the west end of the girders could be hoisted by the crane into position on the abutment bridge seats (similar to what Gulf Concrete did on the Clearwater Bridge). It was a slick operation that greatly reduced the size of the crane required to hoist the east end of those girders, due to the much shorter radius required to reach the trolley. Those approach span girders were 53 m long (174') and weighed 52 tons each. They were so long in fact, that they had to install a temporary steel lattice frame to the top flange to keep them from rolling over due to their length and overall weight. This frame could be removed once the girder was fully braced in position.

Once the west span was erected and we had completed the east pier and abutment, Brittain installed two similar beams on the east pier; however, rather than mount the other end of these beams to a retaining wall, as they had on the west span, they sat that end on timber cribbing near the old highway. This allowed the same erection process to occur on the east span, which again, went very well. The erection of the centre drop-in span was done in a similar manner to the approach spans. The launching beams that were attached to the pier caps were relocated to connect to the end of the approach span girders so the centre span girders could be individually hoisted by two cranes onto the rails and then moved into their final position on the trolleys. It was a very innovative launching system that worked extremely well. Brittain Steel completed their work by the end of January, which, unfortunately, was one month later than originally planned. That only provided us three months to complete the entire deck, which was virtually impossible! (Is this starting to sound like a reality show on TV?)

Fortunately, as we recognized the overall schedule deterioration earlier on in the program, we collectively came up with a great idea to install precast deck panels on top of the girders instead of conventionally forming and stripping the deck, which would potentially shave months off the schedule. Steve McAlister and I started this planning in early October, and after many meetings and working out the required details, we received final approval from the Ministry for this option on November 26, 1985. One of the huge challenges with precast deck panels was figuring out how to accommodate the variable haunches required along every girder to end up with a consistent deck thickness and design profile. Essentially, what we had to do was customize each and every panel to accommodate this variation in thickness, which was a huge logistical undertaking. A contract was awarded to Genstar (Conforce Structures) on January 9, 1986, and the adjusted budget for the revised precast deck was within $5000 of our original budget. And it now looked like we could meet the original contract completion date! As soon as the structural steel was completed, we surveyed the girders so we could calculate the required haunch heights, and they were all calculated by February 25. Conforce commenced fabrication immediately after receiving the haunch heights, and erection of the panels began eight days later on March 5, partially due to the fast process of steam curing the completed concrete panels. The panels were all cast and shipped in the order of the planned erection, so as each truck load arrived, they went in very quickly, with Brittain utilizing their two big cranes and their crew of ironworkers.

Soon after each span of girders was installed, Clem and his carpenter crew got started forming the overhangs and diaphragms, which were not changed as a result of the precast panel option. Leif Rasmussen, Rainer Schoeffel and Dave Ferguson were key carpenters in that crew. By the time the installation of the precast panels was completed, most of the overhang and remaining deck formwork was well underway, as well as the deck rebar placement. Rebar for the project was supplied by Prince George Steel, and their placing foreman was Rick Ross, who was another great guy to work with. A total of 941 m^3 of deck concrete was placed between April 8 and 17 using Cemcon's Schwing concrete boom pump equipped with a splitter to service two deck machines, one running on each three-lane side of the highway that was separated by median rebar down the centre. All concrete was supplied by Hope Ready-Mix. Steve Jarotski ran the Goodbrand deck machine while the other machine was rented from Roger Abbot at Protech Construction, who supplied an operator for it. The main concrete finishers for those deck pours were Earl Hearnstead and Dale (Alki) Richardson, the same guys I had at the Clearwater Bridge four years earlier.

Pouring Deck Concrete – Alexander Creek

One other major time-saving concept that had been under discussion for months was the option of extruding the concrete bridge parapets in order to save time. Normally, bridge parapets would be hand formed using steel or prefabricated wood parapet forms. Assuming that you had 200' of forms, that would require at least eight different pours, and each pour would cycle in about three to four days. Therefore, the time required to place the parapets would be a minimum of 24 work days. Extruding, on the other hand, was completed in about three days. Goodbrand had been experimenting with extruding parapets using a Gomaco curb machine for several months, and I was asked by head office to try and come up with the necessary solutions to various problems they had encountered. The first actual run was on the Culliton Creek Bridge on the Sea to Sky Highway and that went pretty well. Further refinements and changes to the reinforcing steel were made for the Alexander Bridge. The parapets were successfully extruded in late April 1986 without any major issues. They are still there today, and after 35 years they still look great!

Once the parapets and centre median were completed, the overhang formwork was stripped using our rolling overhang scaffold and the parapet railing was erected, the deck drains installed and the concrete finishing continued. By this time, the road grade had been completed up to each end of the bridge, allowing for final paving and installation of the median barriers. Equipment and materials were demobilized and final grading of the slopes below the bridge could get started. We were completely finished for the grand opening on May 16, 1986, when dignitaries, including Premier Bill Bennett, travelled the maiden tour of the brand-new highway, all thanks to a great team of diligent and innovative people.

The building of phase one of the Coquihalla Highway from Hope to Merritt was, unfortunately, a story of significant cost overruns. Much of it, in my opinion, was political due to the very tight timelines required to complete everything by May 16, 1986, in time for the opening of Expo 86 and due to the rush all of the design consultants were in to complete designs in time to get the projects out to tender to achieve the timeline. Following completion, many of the projects had claims for delays and acceleration that amounted to many millions of dollars. The Great Bear Snowshed alone had cost overruns to the tune of about $4–$5 million. The Ladner Creek Bridge actually doubled in length due to poor subsurface conditions on the approaches and the resulting cost was far more than double the contract price. Also, to satisfy the premier that his parade would not be marred, the government literally paid contractors to demobilize site offices and fleets of equipment from the site for opening day, then paid again to re-mobilize everything back to the site to complete the work after the entourage had passed. To me, this was an extremely wasteful use of my tax dollars. However, one aspect of the Alexander Bridge that I am quite proud of is that it actually underran in costs to the Ministry by about $150,000. Harry Sandwith ran a tight ship with costs and there were certainly no give-aways, even though we were good friends off the job. However, there was one time when we were doing a very large excavation on the tricky east slope around the east pier when Harry approached me and asked how we were getting paid for that work. I said that I honestly didn't know as it might have been covered in the unit price for slope grading or it might be incidental to our construction of the access roads. Harry went away and thought about that and later told me to keep track of those costs separately because he thought they were not included in the payment items and he would pay for that as extra work. In fact, that added up to about a $40,000 extra. I think Harry retired shortly after that project, and although we often saw each other socially afterwards, that was the last time we worked together, and I certainly enjoyed that experience. Sadly, Harry passed away on July 3, 2000 at the age of 78 years, and his memorial service was a very emotional day for me.

At the end of the day, we ended up with a job profit on the Alexander Bridge of over $1,000,000 over a total duration of less than 14 months. I would be remiss if I didn't mention at this point that this was by far the biggest team effort I had ever experienced on any project to date. In addition to my own on-site team of Art and Clem, who were incredibly capable, other guys like John Jones, the purchasing agent; Al Beamer, the equipment superintendent; and Kris Thorleifson, the superintendent of the highway contract, were all great to work with and were a big part of the success of the project. Contrary to my earlier stated feelings, Jerry Nauss and Ed Beynon were also very helpful, and it was a pleasure to work within their organization within the limitations that I had arranged early in the project. A significant amount of the written correspondence over the course of the project was actually written by Jerry while most of the meeting minutes about the many serious issues were written by Ed. While I will admit that building this significant bridge project in the middle of a "dirt job" wasn't always easy, one of the pleasures I enjoyed from time to time was driving up to a big rock cut a couple of kilometers away and just watching the production of the big iron moving the blasted rock. In that large rock cut, Goodbrand had one D10 and one D9 dozer pushing the blasted rock down a slope to two 992 loaders loading out a number of Cat 773 rock trucks, and I could sit there taking a break for 15 or 20 minutes, enjoying that massive production. Considering each of these giant loaders could load approximately 8 m^3 every 30 to 60 seconds, that equalled a peak production of about 1200 m^3 per hour, which destroyed that rock cut in pretty short order!

Goodbrand's Big Rock Cut on the Coquihalla Highway Construction
Kris Thorleifson, Superintendent

Notwithstanding Vern Dancy's valuable assistance from head office throughout the work, the most significant contributor by far, was Steve McAlister. Steve worked very hard with me to finish the project on time by helping to solve the biggest problems throughout that project. These problems included working with Devpro to get the caissons installed correctly and quickly; working with Brittain Steel (who was suffering financially at that time) to get the structural steel completed, shipped to the site and erected as quickly as possible; and stickhandling the major design changes for precast deck panels and the ability to extrude the parapets. Unfortunately, that was the last time I ever worked with Steve, which was a shame because we collaborated so well, and we made millions of dollars for the company together.

Notwithstanding the huge success of the Alexander Bridge project, I was given notice of a layoff even before all of the final grading earthworks below the bridge were completed, and I think they had Frank Schultz, who was more senior at Goodbrand, complete that work. I was advised that the reason for my layoff was because Goodbrand was shutting down all operations in B.C. As I recall, they had some interest in continuing to operate in Dallas, Texas and possibly in Alberta. This sudden announcement was shocking to all of the Goodbrand staff and hourly employees as the company had made many millions of dollars on all of their major contracts throughout the province, most recently including the Coquihalla and CP Rail Rogers Pass projects. To me, it seemed like a gambler had just won a huge pot and was instantly cashing in, taking no more bets, regardless of any effects to the dozens of dedicated salaried staff or the hundreds of hourly employees who had contributed to that pot! That was a real shame as Goodbrand had so many talented people with such diverse skill sets and high reputations in the industry. The company was extremely successful, and in my and many others' opinions, it could have been the most successful construction company in B.C.'s history. Many key people from Goodbrand's most successful years went on to form independent companies of their own and became very successful including Bob Burns, Joe Depedrina, Wilf Myron and Gary Koehn, who started Bay Hill Contracting; Bob Charleton, Hugh Whitworth, John Jones, John Wasieczko, Garry and Doug Harper, Norm Knowles, Ray Amundsen and Doug Bloomfield, who started Gemco Construction; Glen Walsh and Darryl Salanski, who started Tercon Construction; Kris Thorleifson and Mike Grant, who started TAG Construction; Ron Amos, Al Goodbrand and Rick Beedle, who started Matcon Civil Constructors; Darryl Zazaluk and Frank Reddenbach, who started Talus Construction; John Gregson and Bob Wallace, who started GCL Contracting; Peter Vanderzalm and Henry Boschman, who started Cap Ventures Construction; and Hans Heringa, who started Sound Contracting on Vancouver Island with Don Van Dam and, I believe, Rocky Ostaffy. Many other key people, including Drew Copley, Ted McCallum, Ron Bruhaug, Jim Mutter, Rick Wagner and Fred Boonstra eventually went on to become senior managers with Lafarge Canada, the company that acquired the aggregate and ready-mix division from Goodbrand. Tom Beck went on to have a very successful career as a senior executive with PCL Canada along with Bob Fouty and Dale Anderson. Ray Pledger and Vern Dancy went on to work with Ledcor Construction after the shutdown.[12]

---

12 I apologize for any others that I have missed in this list.

Steve McAlister had a great interest in placer gold mining for a long time and over the last year or so had spent a great deal of time thinking about that. After Goodbrand closed down he started Talon Fabricators, and although small scale at that time, he developed many copyrighted improvements to basic concentrators and other mining equipment. Steve put all of his energy and efforts into that new business, and he continues to this day to manage a very successful company that has worldwide operations.

CHAPTER 15—
Chilliwack Golf & Country Club

I STARTED PLAYING GOLF AT the Chilliwack Golf & Country Club in 1985 when I moved from our house in Newton to Abbotsford while I was working in Hope, B.C. on the Alexander Bridge. That move allowed me to commute to work daily, whereas the commute from Surrey to Hope every day would have been too much. I could only play on Sunday mornings in those days due to my workload, and a regular group of guys from Goodbrand who played golf got together frequently, including Al Beamer, John Jones, Al Rourke, Kevin Sansom and myself. This was when I first started playing quite regularly and pretty soon, I got down to around a 10 handicap. We used to love getting out on the dewy grass early on Sunday mornings. We were all pretty competitive and enjoyed playing lots of little money games between us. Al in particular was a very competitive guy and playing with him inspired me to play better. On one morning with that group, we were playing in the early spring season and it was pretty cold, but I was playing very well. I think I was even par approaching the 15th green when it started hailing! The greens started turning white and when we putted it created rooster tails, but I would not let anyone quit. I was bound and determined to finish with a low score to post, but the putting became so tough I think I ended up with a 77!

On another round that I will never forget, I was even par sitting on the 16th tee box, when I started to do the math in my head: "Let's say I par 16, because its not too tough a hole, then say I make a good tee shot on 17 and two putt for another par, then make a bogey on 18, which is a pretty long par four. That will be a 72!" Well, needless to say, I was in foreign and very unusual territory with that score and my motor neurons sent by my brain to my muscles ceased to function correctly at that point. I pushed my drive on 16 out of bounds right with a flapping chicken wing left elbow, then overcorrected and hit my third shot off the tee a bit too far left into the tree line, requiring a punch out. I got on the green on my fifth shot and putted two for a triple! Then I was rattled. I continued to bogey the par 3 17th hole and double- bogeyed 18 for a smooth 77. Since that day, I have never liked to add up my front nine score when I am playing really well because it is better not to let your mind start going to places of potential discomfort!

I started taking lessons soon after joining the club and came to really enjoy working on the range with the head pro, Rob Cummings. He was really the first golf coach I'd had, and I absolutely loved the lessons. He was excellent at providing easy references for me to think about or imagine, without filling my brain with minute details of club angles, body positions or grip pressures, etc. He taught me how to swing the club from the inside, generating a draw, whereas pretty well most coaches nowadays will teach you to swing from the outside, creating a consistent fade. On one particular session on the range, I remember him picking targets for me, telling me to hit three or four draws at this target, then three or four fades to that target, and I was hitting the balls so pure and consistent, he thought he was wasting his time and asked me how my putting was. Well, that was my first putting lesson. I was, up until that point, under the misconception that putting was all about feel, rather than some standardized technique. After putting a few balls for him, he declared "We got some work to do here!" It turns out that my head was moving, my body was moving, my knees were moving and my wrists were breaking, apparently all bad things for a good putting stroke. I learned a lot about putting that day, including a lot of drills to improve those skills, and I continue to learn more about putting right up until this day. One thing I know is that you have to think you are a good putter to be one and periodic lessons and appropriate practice can certainly make that belief true.

After playing at the club for a number of years, Rob asked me to become involved in a Long-Term Planning Committee. One of the committee's long-term goals was to build a new clubhouse as the existing facility was old, shabby and beginning to cost too much money to maintain. I worked on that committee for about a year; however, the very conservative Board of Directors was not really engaging with these concepts. Around that time, Rob seemed (to me at least) to become disenchanted, and soon after, he left the club to become a mortgage broker. Rob always took the high road and never told me or anyone else that I know of why he left, but that was a real sad day for the golf club. Needless to say, the old clubhouse remains to this day about 20 years later.

The Chilliwack Open was an annual tournament every July. It was the highlight of the golf club and it soon became the absolute highlight of my golf year. I loved the Chilliwack Open and looked forward to the competition and the additional stress that it brought with it. It included a practice round on Friday and two rounds of medal play on Saturday and Sunday. The course was groomed to its finest condition by Duane Grosart and the rest of the greens crew, and people came from all over the province to play in the event year after year. The tees were set way back and the pin positions were set up to make it as challenging as possible, but not crazy like an ironman competition. On Saturday night, there was a big buffet dinner, a dance with a local band, and lots of crib games going on in the booths around the lounge. You had to be careful Saturday night because if you had too much fun you would be hard-pressed to put together a competitive round on Sunday morning. On one occasion that I still remember, I scored low on Saturday, which provided me a late tee time on Sunday afternoon, so I could let loose a bit on Saturday night, but a late tee time did not help in this case because the effects lasted all day on Sunday, and I found myself out of the competition. However, I did learn from my mistakes, and I eventually won my flight several times in the Open, winning TVs, new drivers, telephone answering machines, etc. On one of those wins in the 1997 Open, my good friend Gary Evans caddied for me, and I played very well, shooting 75 on Saturday with 29 putts and 77 on Sunday with 30 putts, all as an eight handicap (net 67 and 69). In addition to carrying my golf bag and helping me read putts, more importantly, he continually harassed me to "stay in the shot." Before virtually every shot, he said "If you don't stay in this shot, I am going to kick you square in the nuts!" I am not sure whether it was his mantra before every shot or the old supply of Emtec 30 prescription painkillers that I was on to relax my back during those two days, but it sure worked well.[13]

On October 16, 1994, during a round at Chilliwack with my brother Bill, I got a hole-in-one, which is still one of the highlights of my golfing career! (I have not had one since.) Hole #12 was a 187-yard par 3, and I used a four iron from the white tees. After making solid contact and watching the initial ball flight, I said to Bill that "This is going to be close!" Unfortunately, the flag location was tucked just below a rise in front of the green, so we didn't actually see the ball go in the hole. As we walked up to the green, we didn't notice my ball on the green but as we were looking for the ball, Jim Connors, a fellow member, was yelling back from the 13th tee box that the ball went into the hole. It was somewhat less momentous having not seen the ball go in, but nevertheless it was a hole-in-one. Coincidentally, only a week or two later, I eagled[14] the par 4 on the 15th hole, making a shot from 110 yards out, and Jim Connors was on the 16th tee box and also saw that shot go in.

Jim was a very interesting guy who was always well tanned like the actor George Hamilton and also very nicely groomed. One great trick he had was when he came into the clubhouse lounge after a round with his wife, Jessica. Rather than order from the waitress at the table like everyone else, Jim would go up to the bar and order a large glass of beer and a large can of beer for himself. When he set them down on the table, Jessica, who I don't think drank, thought he only had one can of beer, which he had poured into the glass, but whenever Jessica wasn't watching too closely, he continued to top up his glass from the can. Pure genius!

Embarrassingly enough, for some odd reason, I was using an old X-OUT Titleist ball that looked very well used when I got my hole-in-one, so that exact ball still remains in the Hole-in-One Trophy that I received

13 FYI – the lowest scores I ever had at Chilliwack G&CC was 74 but in one of those rounds with Joe Van Essen and Gladys I started with a double bogey and finished with a double bogey as an 8 handicap! Now, I never feel bad starting with a double bogey, specially at Fairview Mountain GC where the number 1 hole is probably the most difficult of all 18.

14 Al Beamer is moaning by now for sure!

from the club! In the old tradition of getting a hole-in-one, many people assembled in the lounge that afternoon for a free drink from the poor sucker who made the memorable shot! I never knew how popular I was at the golf club until that afternoon, and I came to learn that many of my friends loved doubles of Chivas Regal!

CHAPTER 16—

Mother's Day Golf Tour— 34 Years and Still Going[15]

IN THE LATE SPRING OF 1986, a few of us who worked for, or were involved with, Goodbrand Construction on the Coquihalla Highway in Hope, B.C., decided to take off for a weekend of golf in Kamloops. The foursome that travelled that year was Al Beamer, Darryl Salanski, Rod Graham and I. I was a superintendent on the Alexander Bridge, Al was the master mechanic in Hope, Darryl was project engineer and Rod was a salesman for an equipment vendor Al used regularly. We played two rounds at Rivershore and one at Kamloops Golf & Country Club, and we had a hell of a good time. It was a great break from our hectic work life, where we were trying to get the Coquihalla finished in time for the grand opening. Well, the next year, some of us were doing some advanced planning to do the exact same thing and we decided to go the weekend before the May long weekend, so as to avoid the long-weekend rush. Little did we know at the time, that weekend landed on Mother's Day, but not only did we get away with it, we continued on that same date every year and called the tournament the Annual Mother's Day Golf Tournament (MDGT). The group expanded the next year, and we had at least two foursomes and twice as much fun. By about the third year, a group of friends, including Kevin Sansom, who I had gone to high school with and then worked with at

15 Actually, we cancelled the last two tournaments due to the Covid-19 epidemic but with any luck we will be back at the tournament in the spring of 2022.

Goodbrand, joined in from Nelson. In subsequent years, the MDGT kept expanding until we regularly had about 20 to 24 players. We tried to play the best golf courses in an area and maintained the original concept of inviting people related to our construction industry. We played in many great golf destinations over the years, including Kamloops, Nelson, Spokane, Coeur D'Alene, Kelowna, Victoria, Nanaimo, Courtenay, Campbell River, Bend, Oregon, Cranbrook, Kimberley and even San Diego, Phoenix, Laughlin, Nevada, Palm Springs and Mesquite, Utah for special anniversary years.

One of the most memorable trips was to Coeur D'Alene, Idaho in 1992. The resort course there was like nothing most of us had ever seen. There was bent grass all throughout the course, and it was absolutely meticulous. Based on the rest of the course, I believe the floating, moveable island green was way overrated. We stayed at the main resort hotel and even though it was a little on the expensive side, we had a fantastic time. We spent one evening in the hotel lounge and Al Beamer, who I was rooming with on that trip, was buying a lot of drinks on the room tab for some reason. Well, when I woke up the next morning, the bill that had been slid under our hotel room door totalled $682.00 USD! That may or may not have been the same night Willy Stugis got pretty lit up. A number of people were on the dance floor adjacent to where we were sitting at our high-top tables, and I do not recall whether Willy, who was quite a large guy (much bigger then, I think), was attempting to dance or just making his way through the dance floor, but he started to waver and we could see that he was top-heavy and could go down. A number of us were watching his progress in slow motion with trepidation when all of a sudden—WHAM! He went down like a ton of bricks, and unfortunately, a lady who was dancing in the wrong place at the wrong time got absolutely flattened! Several of her girlfriends were screaming and pulling Willy off her, but he was nearly comatose! The place was in an absolute uproar, and I think Willy won the annual Surly Man Award that year. That coveted handmade trophy was presented by the Committee every year to a contestant deserving of that high honour for a particular attitude or act, often (but not always) occurring after imbibing an excessive quantity of alcoholic beverages. Each year, the beneficiary would add some additional decoration to the trophy, so it evolved dramatically over the years. Al Beamer, Warren McLellan and

Willy were the most common recipients until someone's wife allegedly threw it in the garbage, perhaps thinking that it was too vulgar an item to be proudly displayed on the living room mantle.

In 1993, I finally talked Robin McFarlane, our corporate lawyer, into coming with us on that year's event to Kelowna, where we discovered Flashback's Nightclub. Other than the regulars who came every year, we also had Mark Sinclair, Blair Squire and David Surridge join us, which turned out to be one of the wildest Saturday nights in 34 years! Beamer was at his finest, along with Chris Nichols, John Jones, Brian Taylor and Paul Manning, and I have no doubt in my mind that the shooter waitress will remember that night for the rest of her life! I know Robin has never forgotten the scene, and he often brings it up in conversation to this day. He said he had never seen anything like that in his life, and to be more specific, he had never seen any man drink more shooters than Al Beamer did that night! The very last image I have of that evening was Chris Nichols singing old English rugby songs in the middle of the busy street after the nightclub closed (+/- 3:00 AM?) with his pants slipped all the way below his knees. But to our credit, we all played golf the next morning at Kelowna Springs Golf Club, and Rudy collected his annual low gross trophy again along with hundreds of our dollars. Our guest Dave Surridge won low net.

In 1995, on the 10th anniversary of the MDGT, we decided to take our tour down to Phoenix, Arizona. That was a fantastic trip with great warm, sunny weather and fantastic golf courses. The group stayed at a hotel right on Scottsdale Road that was perfectly set up for golf groups, and they served free beer and appetizers every evening from 5:00 PM to 8:00 PM! They couldn't have made any money on our group that week. There was also a nice lounge on the property, so we didn't have to travel very far in the evenings if we didn't want to. There were pool tables and darts in the lounge, so we had plenty of lively games going on. This pub was also decorated with all kinds of large cacti, and one night someone found out that they were actually all made of plastic so that ended up in a cactus fight! You should have seen the fear in the eyes of someone who had a big cactus coming at them!

On one particular night, some of us ventured down to a very large nightclub in Scottsdale called Jets n' Sticks, which also had many pool tables throughout the place. The place was packed as it was located fairly close to Arizona State University. At one point in the evening, I was standing at the bar waiting for the bartender to take my drink order and conversing with a tall blonde girl who was visiting from Israel. She had served in the military there, so our conversation was very interesting. Suddenly, I felt the presence of a fairly solid black man beside me, attempting to wedge himself slightly between the blonde and myself and creating pressure for me to move over to give him way to either the girl or the bartender, which I resisted. My initial resistance indicated that this guy was very hard and muscular under his clothing—there was no give whatsoever. In another moment he disappeared, and then the bartender leaned over to me and asked me if I knew who that guy was. I sarcastically responded, "No, I don't know. Who was it, Lynn Swan?" (For those of you who don't recall, Lynn Swan was a wide receiver with the Pittsburgh Steelers in the 70's and 80's). The bartender then told me that the guy I nudged off the bar was Marcus Allen![16] Yikes!

In 1998, Brian Taylor organized the tournament from his home in Phoenix, Arizona and selected Bend, Oregon for the destination. One of the rounds was to be played in Sun River, which was a 25 km drive away. Maybe Brian was not aware that Bend is located at almost 4000' in elevation and it was only May. It was pretty cool in Bend when we got up that morning, but as we drove to Sun River, the outside temperature indicated in my truck started dropping quickly, and we had bets on how low it would drop. I seem to recall that it went down to the high 30's °F, and the closer we got to Sun River, the more people we saw wearing toques and carrying snowboards! Little did I know, it was a ski resort. Well, when we arrived at Crosswater Golf Course, it was absolutely beautiful, albeit a touch cold. Once we got underway and a few holes into the course, things took a bit of a turn for the worse—it started snowing, and I mean big, heavy snowflakes! By the 4th hole, the snow had accumulated to about 2", we couldn't

16 Marcus Allen was an all-star running back who won the Heisman Trophy while playing for USC in college. He then played in the NFL for 16 seasons, 11 of those years playing for the Oakland Raiders, winning the MVP in 1985 and also the MVP in Super Bowl XVIII. He was the first NFL player to gain more than 10,000 rushing yards and 5,000 receiving yards during his career.

find our balls and you certainly couldn't putt on the greens, so we had to abandon the course in our golf wear. We went back to Bend that day and played another course in the cold rain instead of the snow! We were always a bit skeptical of golfing anywhere near ski resorts in May following that experience, although the same thing almost happened to us again years later at Kimberley.

The MDGT endured for 34 years, with a lot of the same people attending since the beginning, when, unfortunately, in 2020 and 2021, we had to cancel the 35th and 36th annual tournaments due to the COVID-19 pandemic. We are hoping we can resume normal operations in 2022. As the tournament is essentially a net championship based on having fun, since the beginning we've used a special handicapping system that encourages everyone to have fun and put less focus strictly on golfing ability. Basically, if you shot poor rounds on Friday and Saturday, quite possibly due to having too much fun the night before, you could use your gross differential for those days and average those out with your incoming handicap. In other words, if you shot 100 on a course rated 70, your gross differential was 30. For example, if you did that on both Friday and Saturday and came into the tournament with a 10 handicap, then your tournament handicap on Sunday was 23 (70/3). If you could gather your wits about you on Sunday and shoot an 82, you would have a net of 59 and quite likely win the Low Net prize and a pack of money. We also awarded a trophy each year to the Low Gross category but that is usually fought out between Rudy Dorn and Brian Taylor.

There are far too many people who regularly attend the MDGT to mention here; however, a few notable regulars include Al Beamer, who has travelled from as far away as Hibernia, Newfoundland; Denver, Colorado; Roanoke, Virginia; and Phoenix, Arizona; Brian Taylor, who regularly attends from Phoenix, Arizona; and Eric McMillan, who travels from Virginia Beach, Virginia and once even drove across the United States to surprise everyone in Kamloops one year, even though he chose not to play golf with us. Other notable players include my brother Bill, who, along with Al Rourke and Brian Taylor, designed a very elaborate computerized scoring system for the tournament that we still use today.

In addition to the annual Mother's Day tournament, in about 1990, I started attending another annual golf tournament with a group of friends from Langley, many of whom were also former employees of Goodbrand Construction. Some of the regulars on this trip included Bob Burns, Drew Copley, Ted McCallum, Barry Ennis, Dik Munday, Len Gamlin, Brian Atkinson, Norm Clark and on and on. These trips were very similar to the MDGT, and we travelled mostly to San Diego, but in some years, we went to Palm Springs or Escondido and even as far away as Myrtle Beach, South Carolina one year. When we were in Myrtle Beach getting breakfast in the hotel one morning, our group of about 16 players ran into another older group of male golfers, who labelled us "The flat bellies", as opposed to "The round bellies", which they had labelled themselves. Thirty odd years later, we are now all "round bellies"!

This was a great group of guys who were generally good golfers and liked the travel as well as the social comradery and intense competition of the tournaments. Actually, I am not sure what was more intense, the competition or the comradery. On one particular trip to San Diego in about 2000, I got back to my hotel room late one night to find that someone had filled my bed with ice! I heard whispered rumours from Al MacTaggart that Ted McCallum did this deed, so the next day, after golf, I drove all over the greater San Diego area to find a butchered pig's head, which is commonly used by Hispanics as a religious table display, at a marketplace. Regrettably, I could not locate a pig's head due to the high demand that weekend, but I did find a nice, fresh and slimy 15 lb. red snapper that I neatly tucked under Ted's pillow in his room. Unfortunately, just as I was fixing his bed sheets Ted walked into his room, and I was busted. He tackled me on the bed like Mike Singletary of the Chicago Bears, but I was able to escape this potentially precarious predicament without any lasting or serious injury.

This same group of mostly ex-Goodbrand people also travelled to a campsite in Peachland every year, and we joined them for a number of years. This was a great time for us every summer, enjoying these families while boating, skiing and swimming in Okanagan Lake, sharing potluck dinners and going on the occasional golf excursion.

One other golf trip also deserves mention at this point. Annette and I travelled down to Phoenix, Arizona for a short holiday in 1994 to visit Al

Beamer and his new wife, Susan. Al had transferred down there to work for PCL Construction. Al and I did a lot of golfing that week, playing at Superstition Springs, where I got a memorable eagle;[17] Ocotillo Springs; and the TPC Scottsdale, but the grand finale was at Troon North. Playing the TPC Scottsdale was always one of my favourite places to golf because it has such a spectacular layout, and I knew the course so well from religiously watching the annual PGA Phoenix Open. On the morning we played Troon North it was a long drive from Al's house in Chandler all the way up a very busy Pima Road to north Scottsdale before the Pima Freeway was built, and it took us a long time to get there, so we were running a bit late. Once we arrived, Al pulled into the parking lot, told me to go check us in at the pro shop, and he would park the truck and get the clubs onto a golf cart. When I checked in, I was shocked at the price of the green fees at $225 USD each plus cart, considering the exchange rate was about 35% at that time ($607.50 CAN went on my Visa for the two of us). Nevertheless, that was maybe the best golf experience of my life as that course was perfectly manicured and we both played well. At the end of the round, I likely paid Al about ten bucks for our individual Nassau competition, and he thanked me for picking up the green fees![18]

17 Al Beamer will bristle when I start talking about my various eagles!

18 Besides Troon North and TPC Scottsdale, my favourite golf courses I have played over the years include St. Andrews in Scotland, Coeur D'Alene Resort Course in Idaho, Toscana and Andalusia in Palm Springs, Grayhawk in Scottsdale, Links at Crowbush in Prince Edward Island, Cabot Cliffs in Cape Breton, numerous courses in Myrtle Beach and San Diego and my own home course, Fairview Mountain in Oliver. Other memorable low rounds while travelling include shooting a 74 at Monarch Dunes near Santa Barbara, CA; a 76 at Port Fairie Golf Club in SW Victoria in Australia: and 76 at St. Francis Bay Golf Club in South Africa. My next book might be about golfing experiences!

CHAPTER 17—
Western Versatile Construction Corp.

The First Five Years

ONCE GOODBRAND CONSTRUCTION SHUT DOWN in B.C. in June of 1986, I had time to assess my life and career. I was 32 years old with a wife, two kids and a nice house in Abbotsford. I had been working for heavy construction contractors on projects for more than seven years, with six of those years as a superintendent in the field. I had been involved in some very successful projects and worked with some great people, but one thing I knew was that I wanted a piece of the action moving forward and there were greater opportunities coming in the open-shop field. Scanning through the paper for potential jobs one day, I found an ad posted by Manning Construction looking for a senior estimator/project manager. I knew that Manning had just completed some huge projects for CP Rail in the Rogers Pass Tunnel and was very successful with those jobs, so I decided to contact Paul Manning, who was the president of the company. Paul and I were the same age, but he was still going to university when I worked for his father, Dave Manning in 1980. We had a good meeting, and while we both acknowledged that I was really not the type of guy to fill the senior estimator roll, he was interested in my project management experience. We decided to give it a go—I would start estimating some projects and then go run them if we were successful. Manning had just opened a brand-new office building in Langley, so that was not too long of a drive either. Well, it didn't take too long estimating various projects before I became a bit disillusioned because of the fact that Manning was a Building Trades Union company, and we were mostly competing against open shop contractors. My concern in those days was that the only way a union contractor could be low bidder on a bridge was if he forgot to put the cost of the girders in the bid! One day, after a couple of months of that, I sat down with Paul, told him of my frustrations and basically said that I didn't want to continue in this manner forever, but if he ever wanted to go open shop, I would be very interested in working with him. Because there was a gap in meaningful bids coming up, and I was definitely not interested in "bidding exercises", I told him that I would take some time off and golf while he thought about my proposal.

Having the time off was great, and I did golf a fair amount during that break. However, after a few weeks went by, Paul contacted me and said that he was formally resigning from Manning Construction and that we were going to move ahead and set up a new company. He was all in on the concept of open shop, and he had been dealing with Barry Dong, a well-known labour lawyer, working out the necessary details for the legal split from Manning. We also started dealing closely with Robin McFarlane, who would be our corporate lawyer, to get the new company set up and operating. Paul was already making arrangements for bonding and banking as well so we could start bidding projects fairly quickly. That was great news! At that point, Paul was to hold 90% of the company shares and I would hold 10%, which was fine with me—he was the one with family capital. However, we also made another deal on a handshake at that time (an agreement that we honoured up to the very end) that if we were going to be partners, we had to be paid and bonused equally.

The first project we bid was the #5 Road Interchange in Richmond in early September 1986. We bid all of the five bridge structures for that interchange directly to a general contractor called Maximum Contracting, owned by Ike Unger, a long-time North Delta resident. (And a former Annieville Little League Umpire, as I recall.) All of the accompanying roadworks were bid by Maximum, and I bid the bridges in my home office basement in Abbotsford in August 1986. We were unsuccessful with that bid; nevertheless, our name was now out there and we were off to the races.

Soon after, we moved into a second-floor industrial office space in Port Kells in Langley and got settled in with desks, computers. and file cabinets. We hired Carol Shortridge as our secretary. She looked after payroll and all of the company typing (this was well before personal computers when we were able to do that ourselves). Shortly after starting up, we hired Karen Woolf to perform a number of financial and accounting tasks, including accounts payable and accounts receivable. Karen turned out to be the most reliable and accurate employee that any company could wish for. We also purchased the Explorer Software system that Goodbrand Construction wrote and marketed for the construction industry. Manning Construction had previously purchased that system and was using it fairly extensively, so Paul was quite familiar with it. Paul chose the company name, Western Versatile Construction Corp. (WVCC) stressing the importance for the company to be versatile in its pursuits and abilities. Paul advised me on September 9, 1986, that we now had a bonding facility in place with Alta Surety, and Dave Slader, who was with Marsh & McLennan, was now our bonding agent. That was the beginning of a long and mutually successful relationship. Around that same time, we became members of the ICBA (Independent Contractors and Businesses Association), an organization that represented open-shop construction in B.C. The contact we had there was the ICBA president Philip Hochstein. Philip was not only a great leader of this organization; he was also a good friend—a friendship that remains right up until this day.

Now that we were getting our business underway, it was certainly a new learning experience bidding as an open-shop contractor. We could not phone the union companies for prices as we had previously done when working for Manning, Goodbrand or even Quadra a few years back, and we certainly could not contact the union hiring hall for workers. Despite this, the advantages of operating an open shop were tremendous. You see, the whole idea of open-shop construction at that time was not to reduce a tradesman's wages or benefits so that we could operate more cheaply; rather, the purpose was to give the company more flexibility in hiring and what workers could do. In the old Building Trades Union environment, workers were dispatched from local union halls, and we got whoever was on the top of the list, no matter what their skill level was. (Remember, it was the old union mantra that all workers are the same, but that is far from true in reality.) We could sometimes name-request certain tradesmen who had worked for us before, but those name requests were very limited for out-of-town work. Contrast this with the open-shop environment where we could hire whomever we wanted to based on a worker's ability, specific experience, work ethic, reliability, etc. The other big benefit to open-shop hiring was the fact that we could move our workers from one geographical location to another, whereas in the Building Trades Union environment, we had to hire from the closest union hall. For example, if we built up a great crew in Fort St. John and our next project was Cranbrook, we had the ability to transfer all or some of them and pay living allowances, rather than going through the local union hall's hiring list and re-training the crew from square one again. The other great advantage to open shop was the ability for workers to perform multiple duties, such as drive a mixer truck and be a welder, as long as a worker was properly trained and experienced to do both. Conversely, in the old union days, a mixer driver would have to be a member of the teamster's union and the welder would have to be hired from the Operating Engineers, so we would need two men rather than one. The jurisdictional boundaries in the building trades sector created tremendous barriers and inefficiencies that we were finally rid of. All of these changes increased crew productivity and worker contentment tremendously. In regards to wages and benefits, it was our intention to pay new workers similar or slightly reduced hourly wages compared to the old union rates for an initial assessment period, but those workers who showed higher skills or productivity could then be rewarded with higher wages. In many cases, highly productive workers were often paid more than the building trade rates. We offered health and welfare benefits to our employees that were always comparable or better than what the building trades offered. This was called a "merit shop" atmosphere, where workers would actually be paid based on their abilities and production, which provided additional incentive for them to be efficient and productive.

McLeod River Bridge - Whitecourt, Alberta

After doing a little concrete bridge deck overlay job for the City of Langley in October and November, we got our first major project in Whitecourt, Alberta, which was a two-lane highway bridge over the McLeod River. Because much of my own background was as a site superintendent, it was certainly understood that my job was to run this project and Paul's job was to manage the company banking, bonding, accounting, major purchasing and subcontracts from head office, in addition to any further business development and estimating. It turned out on that first real job, as we were both new to this partnership, we had to sort out our individual roles and responsibilities, while maintaining full transparency between us. One example that cropped up fairly quickly was where we both thought we were responsible for formwork design, and when Paul submitted his form plans to me, we had to have a little clarification discussion. I had done extensive formwork design and pour cycle planning at the field level up to that point and, quite frankly, I saw a big difference in Manning Construction's standard timber formwork designs and what I had since seen and used for the last seven years. Therefore, I was quite firm that this was to be my responsibility, although obviously Paul would always get a copy of everything for transparency. We used a collaborative approach at Whitecourt, and we came to a good mutual understanding on that matter.

I called Art Penner on November 6, 1986, and offered him his old job back as my assistant superintendent, this time being paid (as opposed to working for free, as he had offered at Alexander Bridge), which he readily accepted. Incidentally, Art had been working for Quadra Construction installing some skylines for avalanche control on the mountains above the Coquihalla Highway, and they had just completed that work. By mid-November, I also had Karl and Paul Schlauwitz on board as they were all great carpenters whom I had worked with on Conveyor Crossing and the Fort Nelson River Bridge. Karl's other son, Mike, who was a great carpenter, also joined us that fall. One of the local carpenters who was really good was Gord Lakeman, and he worked with us on a number of projects after this job. The client for the McLeod River Bridge was Alberta Transportation, and this was the most difficult aspect of the project for us. Mathew McIntyre was a young, green, yet, eager fellow who was the site representative for Alberta Transportation. Mathew certainly wanted to make an impression with his management by enforcing the specifications to the letter, no matter how impractical his interpretations were.

Mobilizing in mid-November, the bridge had two piers located in the river and two relatively simple abutments. We constructed the river piers through the winter by building gravel access roads to each pier, which allowed Canada Caisson to install the drilled cast-in-place foundation piles. Wood framed cofferdams were then installed in both pier locations so we could place a tremie seal, which allowed dewatering of the footings at the lowest annual flows. The pier shafts could then be constructed conventionally in the dry with dewatering pumps operating within the cofferdams. Square M Contracting, from Red Deer, drove the H-piles for the abutments and our crews followed by forming and placing the abutments. All in all, the substructure construction was relatively uneventful, and we used a mix of local workers and some of our own tradespeople from B.C. However, as it was again the dead of winter, we had to enclose all of the formwork and heat it constantly using propane heaters and hoarding panels like we used in Fort Nelson. It is always a challenge to maintain diesel generators for heating and dewatering up north 24 hours per day, seven days per week in the middle of winter in order to keep things from flooding or freezing.

One interesting aspect of the McLeod River project was the supply of rip rap that was required to protect the riverbanks from erosion. In B.C., you could just find a quarry and drill and blast to produce your own rock product, but in this area of Alberta, there was no rock anywhere other than 200 km away back in the Rockies in the small town of Cadomin where they had quarries. Driving that distance was cost prohibitive due to the trucking costs. The other option, which we learned was fairly common in Alberta, was to locate vendors of fieldstone. Fieldstone is essentially an assortment of rock boulders that have been rolled or plucked off agricultural properties with stone boats over the last hundred years and stockpiled around the perimeter of the fields. The fieldstone vendors collect, process and ship the rock to locations requiring rip rap, which turns out to be quite a business venture in the area. I contacted a few contractors in the area who did this sort of thing, including Richard Wolfram Trucking, and eventually he had a network of truckers picking the fieldstone up and delivering it to the jobsite. As the sizes varied quite a bit in this process, it took some mixing at the site stockpile to get a usable product, but we made it work. Mathew, our veritable inspector, took a picture of a test area of rip rap that was found to be satisfactory in size and gradation, then taped that picture to the underside of his hard hat brim, so he could compare all other areas to the test model for acceptance!

Early in the new year, as the substructure was completed, Central Fabricators started shipping the steel girders to the site and erecting them with cranes off our pier access pads. Before the steelwork was complete, I made a management decision to leave Art in charge of the remainder of the bridge. I gave myself a promotion to general superintendent and returned to head office where I was able to assist Art from there while bidding other jobs with Paul. This included a lone site visit to Bella Coola for a tender for three bridges for the Ministry of Transportation. This was one of the nicest scenic drives in B.C. that I had ever been on. It wasn't until 2012, when I was semi-retired, that I ever got around to doing that drive again.

Once the steelwork was completed in Whitecourt, the deck formwork was installed and the deck placed with Paul Schlauwitz and our local crew, along with some brand-new B.C. employees including Joe Mather, Pete Steiner and Glen Mosley. These new guys were all good men and great carpenters who would work for WVCC for many, many years to come. We rented an Alberta deck finishing machine and the deck and sidewalks went very well with the assistance of Harold Chick, a former Goodbrand employee, who ran the deck finisher. The bridge was fully completed by mid-September 1987. A couple of the tougher aspects about working with Alberta Transportation included their visual acceptance criterion for concrete finishing and their fogging and misting requirements for the weathering structural steel[19]. Both of these specifications were so subjective, the owner could basically decide when it was good enough, or tell you to start all over again, which they had no problem doing, no matter how good the

---

19 Weathering structural steel is a specific grade of steel that allows the surface of steel to corrode and that layer will then provide a layer of resistance to further corrosion and eliminates the requirement for other protective coatings.

finishes looked. We certainly did not get rich on this project, but we made it through a tough test with Alberta Transportation, and that was the last time we worked for them.

Mt. Stephen Snowshed - Field, BC

In the meantime, back in mid-April in head office, we were fully engaged with CP Rail for the construction of the Mt. Stephen Snowshed. Larry McKee was a senior construction manager with CP Rail who was in charge of this project from their office in Calgary, and we developed an excellent relationship with him that would last for many years to come. Hugh Robinson and Al Arnold were also project managers with CP Rail that were involved in this project. The snowshed project was a very interesting and challenging undertaking that involved redirecting a huge snow and debris slide path over the CP mainline just east of Field, B.C., where erosion had been a big problem for the railway for the last 107 years since the rail line first opened. The earthwork component of the project required a huge trench to be excavated right from the foot of the glacier near the top of the mountain down the steep incline and earth berms to be constructed alongside the trench at least 80' high! Then, a large precast concrete tunnel was erected where the rail line would be relocated beneath the realigned snow and debris path. The earthworks started at the top of the mountain with a local subcontractor from Golden called J-TECH (Joe Alexander and Jerry Code) who employed a Cat D9, a Cat D8, a Cat D7 and a Cat 235 Excavator as well as a vibratory roller. They simply started dozing material out to the berm location and compacted it while they deepened the trench and trimmed the slopes with the excavator. As they came further down the mountain, they moved massive quantities of soil and rock material with steady progress, moving about 2000–3000 m^3 per day.

In the meantime, we hired Ron Lyttle to be the project manager for this challenging job because he actually lived in Field, B.C. and had been a field engineer for Manning Construction at the Rogers Pass Tunnel. Tony Vandervegte, who I worked with back at Goodbrand, was hired as the superintendent. Tony was a very experienced concrete guy and an easy, likable guy to work with. Rob Schibli was hired out of BCIT as a surveyor for the project.

The first item of work to be undertaken for the snowshed itself was to densify the existing ground material below the location of the new precast concrete snowshed. This existing rail bed had been used for the last 100 odd years with many snow and mud slides destroying the railway, leaving a lot of unknown materials in that area, including old rail trestles and maybe even rail cars! The chosen method to compact this area was to use 'dynamic compaction', which is a fancy name for picking up a huge weight high into the air with a big crane and letting it fall repeatedly, pounding violently into the ground. Geopack (Nelson Beatty) was the subcontractor for this work, and the method seemed ideal for this type of varied ground conditions. Once that compaction work was completed several weeks later, we began our concrete work, building the large foundation footings and head walls for the precast tunnel. As I recall, we rented a concrete batch plant from Max Helmer in Invermere and set it up on site, with Paul Schlauwitz running that plant. We awarded a big contract for the supply of the precast concrete walls and roof beams to Conforce Structures. All precast concrete was delivered by rail and offloaded right on site with our big rented erection crane. At that point, Joe Mather, Pete Steiner and Glen Mosely were finished in Whitecourt, so they came over to assist with the concrete work.

Most of these regulars lived together with their wives in a rented house on an acreage just outside of town that they dubbed "The 9.9 Ranch". There were many late-night campfires at the "9.9", and one night, I elected to stay over for the night rather than head back to my motel in Golden. I spent quite a bit of time on this project and recall staying in a motel in Golden for one stretch of seven days in a row and ordering the exact same pizza every night for dinner that week. Obviously, it was great pizza!

As soon as the cast-in-place concrete foundations were complete, we started erecting the big precast wall panels and bracing them into position. One day, around August 1, 1987, I was on site when a sudden heavy summer rainstorm hit the site. It had been a very hot day and the resulting afternoon thunderstorm and heavy rain was rapidly melting the glacier above and saturating the entire mountainside, causing mud to run down the partially built chute. It began as a wall of mud and very quickly got worse. It was so intense that it was very likely going to wipe out the concrete foundations and the partially erected precast wall, as well as the CP mainline. We made a quick field decision to stack some of the huge precast wall panels with the erection crane and build a temporary buttress to protect our completed work, knowing that would direct the mudflow right across the mainline. But we also knew we had no other choice. Within one hour, 100' of the mainline was completely wiped out and the mudflows continued further down the mountain, completely blocking off the Trans Canada Highway and partially blocking the Kickinghorse River! That was certainly not a great day to be on site. Many hours later, we got one lane opened up on the highway using some of our equipment on site as well as equipment from the highway's maintenance department. CP Rail mobilized their own repair crews and quickly re-established their rail line through the site. It was definitely not the first time this had happened at this treacherous location, but some said this was the worst case they had ever seen or heard of. Luckily, CP Rail was self-insured and they took care of everything. I recall we were fully paid to reinstate the site as well as for our efforts to mitigate the damage and delays.

As soon as the site was cleaned up and the railway reinstated, we completed the precast erection and finished all of the cast-in-place concrete headwalls. Tony, Joe and the rest of the carpenter team produced some very low man-hour productions on that work. J-TEC finished the earthworks over the completed shed by the end of October 1987, and luckily, we got off site before the winter snows began and snowslides started coming down the brand-new chute. All in all, it was a great job and CP Rail was very happy with our work. It was the beginning of a very long and profitable relationship with Larry McKee and CP Rail. J-TEC did a great job on the trench and containment berms under very difficult conditions, and their work remains visible from the Trans Canada today, though unfortunately, we never had the opportunity to work with them again.

The Independent Employees Association

While the Mt. Stephen Snowshed was underway and the McLeod River Bridge was being finished up, several of the company's steady employees decided to withdraw from the carpenter's union and set up an alternative labour union in order negotiate a collective agreement directly with WVCC. It was going to be similar to the alternative Christian Labour Association of Canada (CLAC) labour union agreement that was becoming more common throughout B.C. and Alberta. This move was precipitated by various threats the workers were receiving from union halls stating that they intended to unionize WVCC, and the workers saw no personal benefit in allowing that to happen. As it was, key people like Joe Mather, Pete Steiner, Karl Schlauwitz, Paul Schlauwitz and Glen Mosely believed that as long as WVCC remained competitive and got contracts, it was better to work steady for our company. After all, they could work all over Western Canada and receive good pay, a tax-free living allowance and full health and welfare benefits, rather than work in the Building Trades Union environment where they would be laid off after every job and work for a number of different companies on each successive project. These guys worked with a labour lawyer to set things up, and they negotiated their first collective agreement with WVCC in the spring of 1987. This union was called the Independent Employees Association (IEA). I believe Glen Mosely was the first president of the union, and he stayed on in the union executive for many years. The IEA represented all hourly trades for WVCC right up to the end of the company's operations in 2012.

Fibreco Pulp Mill - Fort St. John, BC

In the summer of 1987, Philip Hochstein, who was the president of the ICBA, approached Paul Manning and asked if WVCC would be interested in bidding some work on a new pulp mill being built in Taylor, B.C. because the owner expressed an explicit interest in open-shop contractors. Of course, Paul was totally interested as this looked like a great chance to get WVCC into some industrial work in the open-shop sector. My first thought was, "Oh my God, do we know how to build pulp mills?" The first bid was for the concrete foundations for the Effluent Treatment Plant, and in September, WVCC was successful in negotiating that contract for the amount of $717,000. Over the next 18 months, we undertook four additional contracts with a total value of $4.4 million, essentially completing the entire treatment plant facility, including all the mechanical and piping, all of the concrete work in the main CTMP (Chemi-Thermo-Mechanical Pulping process) building and a complete rail car dumper facility that would pick up loaded rail cars and dump the wood chips to be used in the pulp processing. We hired Bob Aspinall, a former Manning Construction supervisor from Rogers Pass to be the project manager of the CTMP project. Bill Fielding was put in charge of running the mechanical works while Karl Schlauwitz and his son Paul took care of the initial treatment plant contract. Art Penner went up to the project to assist with the CTMP as soon as he was finished at Whitecourt, and Joe Mather, Pete Steiner, Glen Mosely and Rob Shibli mobilized to the site once the Mt. Stephen snowshed was complete. Tony Vandervegte also came up a bit later to run the chip-handling dumper pit with Karl once the treatment plant work was completed. This was a big project for us, and we needed all hands on deck.

HA Simons was the owner's consultant and responsible for design and project management. Harold Helm was the project manager, and Cathall Fox was the construction manager. Both these individuals were top-notch professionals and HA Simons was a fantastic company to work with. They always knew exactly what we were doing at any time and exactly when they needed to get last-minute design changes to us in order to avoid serious interruptions or delays. They were easy to get along with and very proficient in managing our contracts. As these projects were very much fast-tracked, design changes were constant due to the many details that evolved during their own procurement of process equipment and issues with related shop drawings. Notwithstanding the multitude of changes and constant increase in the scope of work, it was so well orchestrated that we still completed on schedule. For example, when we started the work on the CTMP, the drawings were large, clean, white sheets of paper that were easy to read and understand, but when all the design changes were made, the sheets were literally black with details. I am not sure if I could have possibly bid the work on the completed drawings because they were so complex!

Despite HA Simons great orchestration of the evolving designs, the work was not easy. The CTMP building was four separate floors of extremely complex and interconnected work with literally thousands of anchor bolts, hundreds of pipe sleeves and many dozens of equipment pads, so we assigned one general foreman per floor. We were definitely put through our paces: both our staff and tradespeople had to evolve quickly and accordingly, with increased supervision oversight, quality control and persistent survey checks as the work proceeded, which became part of our company policies on similar projects to come. I was flying back and forth on an almost biweekly schedule to stay on top of the million details that had to be dealt with. This was a very hard job on Bob Aspinall due to the stresses related to the thousands of details and the fast-tracked schedule pushed by the owner.

Paul Manning and I were together on one of those PWA flights to Fort St. John and we were just about to make a stopover landing in Prince George, when I knew something was wrong. While descending in torrential rain and limited visibility, the 737 seemed to be searching for the runway, descending, then levelling off and turning right, then turning left, then speeding up and seemingly ascending in altitude. After a number of minutes, the plane went into its final approach—wheels lowered, flaps down and dropping steadily, when all of a sudden, the plane aborted the landing and accelerated up sharply. The flight attendant announced that the landing was aborted, and we would be making a turn and coming in again for the landing in a couple of minutes. After turning around, we went through the exact same pattern of searching up and down, left and right, then down in a final approach and another sudden aborted landing. This was very unusual and quite concerning considering the bad weather in the area! Ten minutes later, we made a direct, normal landing approach and landed safely at Prince George; however, the delay at the airport was considerable, and we were eventually told to get off the plane and wait to reboard a while later. We took advantage of that leisure time to enjoy a couple of single malt scotches in the luxurious airport lounge. After a couple of hours, we reboarded and made it to Fort St. John without further incident. We found out afterwards that the pilot had twice attempted to land the plane on a radio beacon located on a mountain top and the low-level warning alarm went off in both attempts and caused the pilots to abort the landings! We also learned that the low-level warnings went off at a distance of 75' from the tree tops! The captain was fired on the spot after landing in Prince George and the co-pilot was suspended for six months after the final landing at Fort St. John. Our long delay at Prince George allowed the PWA time to get another pilot into the cockpit to fly on to Fort St. John. Additional details about that flight came to Paul and I some months later while golfing at Shaughnessy Golf Course (where Paul was a member) with some pilots. They confirmed that the co-pilot on that flight had switched the navigation system from one seat to another[20], so the pilot was actually flying to the navigational beacon not the airport! Clearly, that would have been a significant disaster as well as the end of WVCC!

As the Fibreco Mill was the very first pulp mill being built in B.C. by open-shop forces, the building trades were angry because they were losing this industrial bastion they had established and maintained for so many years. They were so angry, they exerted significant pressure on tradesmen employed on our project and seriously harassed some of our employees in an attempt to certify them to the Building Trades Union. In at least one instance, a member of the IEA on our project workforce was questioned in a bar by a union organizer who was wearing a secret recording device and hoping to get damaging information on the formation of the IEA. At least two separate unions attempted to overturn the collective agreements between WVCC and the IEA by making court challenges to the Labour Relations Board. Luckily, all those legal attempts failed, thereby strengthening the legal standing of the IEA. We owe a lot of credit to Barry Dong, our labour lawyer, for effectively managing these legal challenges, along with Grant Gayman, who acted on behalf of the IEA.

When our work on the pulp mill was finally complete, Harold Helm called Paul Manning and congratulated our company "for quality product, professional honesty and reliable performance throughout this project", but he also said that we did not charge nearly enough! I wish he had said something about that a lot sooner as I certainly could have fixed that problem! At the end of the Fibreco Project, WVCC had completed a total of $8,590,000 worth of work over two and a half years, and while things

---

20 As it was described to us by these pilots, this particular Boeing 737 plane had recently been converted from a cargo plane to a passenger plane and that allowed switching the instrumentation between the pilots, whereas this could not usually occur on a typical 737 passenger plane.

were admittedly hectic, they rolled along nicely for a new company. We found great supervisors, mainly from past work, and we also developed a great number of skilled and dependable tradespeople on that project.

Fibreco Pulp Mill

Premier Gold Mine - Stewart, BC

While the Fibreco civil work was well underway in 1988, along with additional mechanical and piping contracts that we were engaged in, we began negotiating some work with Westmin Resources for a new gold mine called Premier Gold that was to be built near Stewart, B.C. Our proposal included all of the concrete work for a new concentrator building, a reclaim tunnel, coarse ore foundations and a crusher building as well as a fair amount of local excavation and backfilling. We were awarded the contract in early 1988, and mobilized to the site in April. Merit Consultants was the construction manager for Westmin Resources. Tony Mochinski was Merit's president, Allan Ahl was the construction manager on site and Joe Rokosh worked as project manager. Tony, who was a former Commonwealth Construction guy, had worked with Dave Manning in the past on industrial projects, and Allan Ahl was a very experienced construction manager from Northern Construction. They all seemed to like working with us (after working out a few kinks in the beginning), so in the end, we performed about 90% of the civil work at the site. Clem Buettner returned as superintendent for us on this project, along with Art Penner as his assistant, Joe Mather working as the crusher foreman and Pete Steiner as the concentrator foreman. By now, these four characters were becoming quite a team, working very well together as long as no one took Clem's constant teasing and sarcasm to heart. Dan Cave was also hired to perform survey and assist with supervision once he graduated from BCIT in May.

One thing I learned from working with Joe Mather over the last few years is that he would do anything as a general foreman to help the company or his men, including aiding or lending a hand to anyone and potentially stretching himself a bit thin. On the Fibreco job, we assigned Joe to be the general foreman, requiring him to run all over the project site from one floor to another assisting wherever he could; however, it may have been better to have him stick to a specific area and delegate the responsibilities for those other areas, so that is exactly what I had in mind for the Premier job. I wanted Joe to look after the crusher building and Pete to be in charge of the concentrator building. The crusher was the most critical piece of the work as the owner required it by a specific date and it was a challenging structure about 70' in height. I personally designed the formwork for the crusher but more importantly, I designed the lift sequencing and planned reuse of the large panel forms along with Clem for an achievable, but absolutely required schedule. Next, I sat down with Joe with the labour budget and the lift and pour sequencing and worked out the average crew sizing required to achieve both budget and schedule. It all looked doable on paper. Next, I provided additional incentive by offering Joe and his wife Marion a free trip to Hawaii if he achieved the schedule and labour budget.

Well, it worked out very well, and Joe, who gained the nickname "Crusher Joe", went to Hawaii on the company. He completed the crusher on time and on budget, and he was very deserving of the reward. I wish more of our work over the years went that smoothly for both budget and schedule!

Premier Gold Mine - Crusher Building

I should mention that the town of Stewart holds records for massive snowfall and there we were again, working right through the winter, snow clearing, shovelling, hoarding and heating. By that point, many in the company were wondering where "Western Versatile" came from as it should have been "Northern Versatile"! It was a task getting through that winter because it snowed several feet per day for many days in a row. Pete Steiner and his crew did an excellent job on the concentrator and the overall team did extremely well on the project. The owner was ultimately happy with our work, with the exception of a little bit of friction between us as we were trying to get started. The job admittedly didn't get off to the best start due to delays, including handing over areas of the site that were being blasted directly for Merit by Edco Contracting and receiving design drawings early enough to do any meaningful planning. I know that during that project I was constantly looking to hire more tradespeople throughout the entire schedule as we could never seem to get enough workers on the site. In the end though, we brought back many people from the past including Glen Mosley, Wayne Clark, Emery Baker, Rainer Schoeffel, Harold Chick and Gord Lakeman. In addition, we gained many more skilled and dedicated tradesmen on this job including Blaine Benny, Nick Maskery, Jerry McMillan, Hugh Wilson, Brad Johnson, Gus Gudmundson, Joe Gardiner and Bob Ramsay. In the end, we completed our contract work by February 1989 and did a total of $3.8 million worth of work on the Premier Mine. This added up to a volume of over $12 million in the just over two years since we started the company; however, that was nothing compared to the next year in 1989, when we did almost $19.5 million in volume!

I visited the Premier site many times, flying into Terrace, renting a brand-new Nissan Pathfinder from the local car dealership and driving at high speed up to Stewart, a distance of just over 300 km. I either stayed in the camp at the site or, later on, at one of the motels in town. The area around Stewart was remarkably beautiful if it was not totally socked in and raining, which seemed to be 80% of the time—although it was remarkably nice and sunny on all of my site visits. Some of the local highlights I experienced on these visits included the beautiful Bear River Glacier and the drive up to the Salmon Glacier where we saw huge chunks of ice the size of houses, which had broken off the glacier above and would remain in the valley through the entire summer. We also stopped along Fish Creek where we saw bears fishing in the river on one occasion. On another trip, I brought my son, Derek, and we went trout fishing with Art up to Little Bob Quinn Lake, north on Highway 37. There was certainly no shortage of fish there, and as I recall, we were the only ones on the lake.

1988 was a good year and a bad year. My mother died on October 23 at the age of 64 years due to a prolonged battle with lung cancer, which was very sad, but my youngest daughter, Chaylene, had been born in Abbotsford on June 9 of that same year, which was a very happy change to our lives.[21] As for WVCC, we were booming along with our revenues increasing, profits rising and labour and equipment resources expanding. But nothing came easy. My job was extremely challenging, and managing the expanding workload was becoming increasingly difficult. My personal responsibility had evolved over time to involve managing all of the civil field operations, and as I am considered a "detail person", I ended up with exponentially more details and problems to manage as we increased scope and volume. Based on past experiences, one premise of our company operation was that it was not feasible for one person to be totally committed to estimating and construction management at the same time. Therefore, because Paul oversaw all estimating, he couldn't provide much construction management assistance to reduce my workload. Others came and went to oversee our fluctuating mechanical operations, which ultimately fell under Paul's responsibilities, but the vast majority of our work was civil. Vern Dancy, the former civil estimator from Goodbrand, joined our team in 1988 as senior estimator, which greatly expanded our estimating capacity. Vern was a guy I worked very well with at Goodbrand, and although he was very helpful to me at times, his primary responsibility was estimating, so his ability to assist on an ongoing basis was somewhat limited.

My brother Bill, who had a degree in finance and industrial accounting from UBC, also joined our company as our comptroller a bit later in 1991, so he took over all accounting, financials, banking operations, and information technology (IT). His previous employment was as a controller of a large manufacturing and distribution company, and his most recent step up that ladder took him to Toronto, but he could not face living there for the rest of his life. Bill easily handled all of our typical accounting and financial operations, and he also dove into the Explorer Accounting system, including all of the functions it encompassed, such as job costing, equipment maintenance and equipment repairs. His involvement in costing and equipment maintenance was another big advantage to me because this

21 The birth of our last child, Chaylene, was to also be the end of my habit of smoking cigars. Although I had quit many times over the previous 10 years or so, every time I had another child, I celebrated with the old-fashioned custom of handing out cigars, and that was it for me, all over again! My favourite were the rum-dipped, wine-flavoured Colts but after quitting that time, I never smoked again.

freed me up to deal with the more direct project management functions. Bill became a huge benefit to the company over the years as he took on more responsibilities from year to year. One of the tasks he was involved with a few years later, along with Marny Woolf, was the creation of an estimating manual, which was based on all achieved costs and productivities to date. This put our costing data to very good use and provided guidelines to anyone doing estimating work for us. Bill also became very involved in the management of the Manning Family Trust, which involved developing industrial land and managing buildings owned by the Trust.

Hugh Brown, the former superintendent from Manning who I had previously worked with in Lillooet, joined us in the early days of WVCC to assist us in the shop and yard as well as in purchasing and managing the tools, equipment and material we were beginning to amass. In the early days, it was a full team effort to load out trucks in the yard that were going to the sites, and we all had to be adept at running the forklift and the 20-ton Pettibone hydraulic yard crane. Hugh was an excellent fit for this position because of his vast experience in our type of construction. He worked for us a number of years and then John Hope was hired part-time in September 1992 to replace Hugh when he eventually retired. John was also a great fit because he had hands-on experience working with Port Kells Equipment Rentals and he was a very successful local farmer who also had spare time to help us whenever it was required. John took full ownership of the shop and yard, including a lot of the purchasing and equipment maintenance functions and ended up working with us for 20 years.

In 1989, WVCC took on nine projects with a total value of $19.5 million—four of those being fairly minor and adding up to about $800,000. Those smaller jobs included a water treatment facility built in Sechelt by Tony Vandervegte with Dan Cave assisting; a welded steel gas pipeline crossing the Skeena River in Terrace for Pacific Northern Gas, which used our modified bridge deck stripping scaffold from Whitecourt, completed by Clem, Art and Joe; some mechanical work by Bill Fielding at Samatosum Mine near Barriere, B.C.; and a concrete stairway for the City of Red Deer by Pete Steiner. The other five projects in 1989 were fairly substantial contracts and three of those were bridge projects in Alberta.

Red Deer River Bridge

The Red Deer River Bridge was a new railway bridge we were awarded from CP Rail in 1989 in the amount of $3.5 million. The substructure of that bridge was supervised by Don Tasker and Pete Steiner. Don had previously worked with Kenyon Contracting for many years as a general superintendent and was very experienced in bridge construction. We worked together briefly while I was in Kelowna with M.O.T.H. and Arnold Talbot in 1979. The bridge was necessitated by a rerouting of the CP rail line around the City of Red Deer, Alberta and included two piers in the middle of the river as well as two concrete abutments. The river piers required sheet pile cofferdams that we installed ourselves with a rented vibro hammer, and all four substructure elements required foundation H-piles driven by Great Plains Piledriving. All of the concrete work was self-performed. The structural steel was supplied by Dominion Bridge in Winnipeg, shipped to Red Deer by CP Rail and then trans-shipped to the site by Premay Equipment, who were heavy-hauling specialists. WVCC co-designed an erection scheme for the structural steel with Peter Saunderson and Jeff Mullins at Somerset Engineering. We rented a bare 250-ton Link Belt crawler crane from GWIL Cranes and used a small crew of ironworkers supplied by UMA Spantec to erect the steelwork. This work was completed in December of that year, and the remaining deck forming, concrete, railing and finishing touches were completed by Dan Cave in the spring and summer of 1990 after Don retired and Pete had moved on to the Ashcroft Bridge. John Jaconen, Nick Maskery and a few others from the Pincher Creek bridges worked on the deck with Dan.

Road Link H - Pincher Creek, Alberta

The next significant project that year was Road Link H Bridge near Pincher Creek, Alberta, which was a two-lane highway bridge worth $2.5 million that was tendered by Alberta Public Works (not Alberta Transportation) in March 1989. This was a new bridge over a valley that was to be filled with water as a result of the completion of the Old Man River Dam. That bridge had three fairly high concrete piers, two fairly simple abutments and a steel plate girder superstructure with a cast-in-place concrete deck. This project was run by KC Shenton, a good local guy who had a lot of experience in bridge and heavy construction and plenty of local tradespeople to draw from. Dan Cave was brought down to be KC's assistant superintendent and surveyor. That bridge was attractive to me because we knew we had a very good chance of getting the much bigger Castle River Bridge project that was close by and out for tender from CP Rail. Road Link H was one of the few times that I personally bid a project and got the job. It is probably true that if you are too detailed a person, you may not be low bidder very often, but it has always bothered me to think that you couldn't get a job if you did too thorough a job of estimating.

DeGraaf Excavating, from Lethbridge, performed all the mass excavation and bridge end fills for this project. One clear recollection I have of the start-up of that job was driving all over Southern Alberta with Pete DeGraaf in his pickup truck on gravel country roads at very high speeds looking for fieldstone to be used as rip rap, as we had done in Whitecourt several years earlier. The speed we were travelling was terrifying, but because everyone could see the vehicle coming for miles due to the huge dust plume, Pete considered stop signs redundant! Other than that scary day, Pete DeGraaf and Al Snow were great guys to work with on that job.

KC and his mostly local crew, including Vince Anderson, Ken Dobbs, Roy Trodden and even Emery Baker from Avola, built the substructure with a rented GWIL 70-ton crawler crane that was operated by Ken "Shakey" Arthur with his small terrier dog at his side. One incident I was reminded of by KC during the writing of this book occurred on the substructure when Dan Cave was performing some survey on the bridge site and the theodolite blew over in the wind, causing serious damage to it. (Oh, yes, it gets very windy in that country.) Apparently, after that Dan never heard the end of it from me about the error of his ways for leaving a theodolite unattended, which was breaking a cardinal rule in the business, so I might as well mention it here one more time.

Scott Steel was awarded a subcontract for the supply and installation of the structural steel for Road Link H, and they were ready to go in time to utilize our rented 250-ton GWIL crane that had just finished erecting girders at the Red Deer bridge. Again, Somerset Engineering managed the design of the erection scheme. Once the steel was completed, KC and his crew got busy on the deck formwork, which proved to be slightly different than previous decks we had built due to the very frequent high winds in the Pincher Creek area. These forms actually had to be designed for hurricane force winds with special tiedowns so they wouldn't blow away! On a side note, on one particular day while wrapping up that bridge, KC noted in his daily report for the day that it was so windy the round bales of hay in the adjacent field all rolled to the fence line! He confirmed this was a true story during the writing of this book. There were many days when it was impossible to work on the deck because of high winds and many of the crew would congregate at the Lundbreck Hotel tavern, where the local RCMP was required to encourage this unruly group to disband on more than one occasion. The deck concrete and bridge completion went very well under KC's supervision, along with the help of his crew of men, including John Jackonen and John Makreel, using a deck finishing machine rented from Getkate Construction who were based in Alberta.

Road Link H Bridge

Castle River Bridge - Pincher Creek, Alberta

The third bridge we were awarded that year was the Castle River Bridge for CP Rail, which was also required to carry the railway across the valley that would be flooded as a result of the completion of the Old Man River Dam. The tender value of this project was $8.1 million, and it was a major structure, far bigger than the previous two, with three massive river piers and two shore piers, as well as two abutments. Clem had finished the gas line in Terrace and he came down to Pincher Creek to run this job, and Dan Cave came over from Road Link H to be Clem's assistant superintendent. Richard Evans, also a newly hired BCIT grad, was hired to go assist KC on Road Link H. Gord Stager, who was a contact of Joe Mather's from Cranbrook, was Clem's carpenter foreman. Top Notch Construction had a large subcontract to construct the substantial bridge end fills, and they were a great company to work with. Alex Lockton, who in my memory looked something like Hoss Cartwright from Bonanza, was their president, and those tobacco-chewin' cowboys knew how to move dirt—and quickly! Most of the dirt was moved by 631 Cat scrapers and D9 dozers. Tony Bowman was the site engineer for CP Rail, and we found him great to work with because of his experience and great practical sense. He and Clem worked seamlessly throughout this project.

Due to the extensive pier construction on the Castle River Bridge, we decided to purchase a crane rather than rent another one because we could easily keep it busy with all the bridge and industrial mine work we were getting in those days. I eventually found a 60-ton Koehring conventional truck crane for sale that had recently been completely rebuilt and restored to near original condition by Cana Construction in Calgary, who had been its original owner 23 years earlier. We bought the crane on June 1, 1989, for $75,000. Cana also told me about the crane operator who had run the crane since it was first picked up from the dealer in 1966 and suggested he might be interested in continuing to run it for us. I telephoned the fellow, Lloyd McHarg, in Calgary and he said he would be ready to go on Monday morning. Lloyd came with me to Cana where we loaded all of the crane components and parts and shipped them to Pincher Creek. It was clear he knew that crane backwards and forwards. The crane turned out to be a little oddball with lots of quirky things like a gas upper engine with down draft carburetors and air over hydraulic operations, so it was a good thing that Lloyd came on board. That crane worked out perfect for the piers at Castle River, and Lloyd and the crane stayed on with us for many, many years. Many years later, we converted the upper engine to a Jimmy diesel, which was a lot more reliable, but it was loud and it spewed oil, which I understand Jimmy's were famous for. Because the crane was an oddball Koehring make, whenever repairs were necessary, we often had to have the parts custom made in machine shops; however, we always made it work thanks in part to Pat Levigne, a very experienced crane mechanic from Brittain Steel, and later Northwest Piledriving, which was owned by my old friends Colin Corbett and Mike Pritchard.

There was about 10,000 m^3 of concrete on the Castle River Bridge compared to Road Link H, which had about 3,500 m^3. Initially, we intended to utilize the local ready-mix company (which I recall was Glen Bros. Trucking) for all of the concrete supply because they provided a competitive quote for concrete on Road Link H; however, when they quoted their price for the Castle River Bridge, it went up from ~\$85.00/m^3 (Road Link H) to ~\$105.00/m^3. This made no sense to me as the increased volume should have provided even better economics. When we received this quotation, I called Don Glen, the owner, and told him we could not accept that price as it would be cheaper to mobilize a concrete plant to the site and produce our own concrete. I told him to rethink his pricing because we were happy to award the concrete supply to him at a reasonable price and then to let me know because otherwise, we would go it on our own. Unfortunately for Don, he held firm with his inflated pricing, and we had no choice but to produce our own concrete. We went into action very quickly. Paul Manning got started by purchasing a batch plant from Manning Construction that had been used in the Rogers Pass CP Rail Tunnel for several years and was now sitting idle. He and I were also looking to purchase five or six used mixer trucks. In the meantime, KC found a good location in Cowley, Alberta with a paved yard, a good supply of water and a shop to rent from Ken Burnett, where we would set up the plant. Bill Fielding got into action mobilizing the plant to Cowley and made a number of significant modifications to make it better suited to this operation. We started looking for suitable aggregates locally, when we hit the gold mine! Locating suitable concrete aggregates in Southern Alberta can be very dicey because of alkali reactivity in many areas, which can cause a lot of serious deterioration problems in the concrete over time. The potential pit location we found was basically a big field of grassland near the Castle River that was never used for aggregate production before, so I had to expedite the necessary petrographic analysis and other CSA tests to determine if the aggregates were acceptable. All of the tests came back 100%, so we quickly retained MP Crushing out of Lethbridge to produce the required aggregates, which they did for about $6.75/TE, delivered to the plant. That was a huge savings from the only competitive alternative, which was hauling all of the aggregates from Reg O'Sullivan in Fort McLeod, about 40 miles away. I also negotiated a cement contract with two options: one from Lafarge, a company we had worked with many times in the past (and I really enjoyed working and golfing with Jim Pearce over the years) and the other from Ideal Cement, a company based out of Montana, which wasn't that far away. The savings to go with Ideal, however, was $36.50/TE and with the 7000 tonnes that we would eventually purchase, this was a clear savings of over $250,000! We eventually purchased six mixer trucks and mobilized the old 966 C Cat loader[22] that we'd bought from Manning Construction (the same one that John Findlay operated in Lillooet!) Ken Murray, one of the batch plant operators for Manning at Rogers Pass, was hired to manage the batch plant along with Gary Gallagher, who also worked in Rogers Pass and took over from Ken after a short while. In order to deal with the hot weather in the summer, we purchased a gas-fired water chiller that would chill 4000 gallons of water from 50°F to 36°F overnight to keep the concrete temperature as low as possible. This was very important for some of the big pier pours on the Castle River Bridge and prevented us from having to use bagged ice to cool the fresh ready-mix concrete, which would have been much more costly and labour intensive. We also purchased a big old steam boiler from Manning Construction (manufactured in 1952 according to Gary Gallagher) to deal with winter concrete when necessary. Once we were all finished the two bridges, including additional local sales to other contractors, we produced 22,499 m^3 of concrete with this operation. The unit cost was $97.10, including the total cost of mobilizing the plant and making all the necessary modifications to the plant, as well as the significant amount

22 The old Cat 966C loader had a long life of its own that was quite remarkable. Manning Construction bought the loader new in 1970 and used it all over the province on many of its jobs, including the Lillooet Bridge. WVCC bought the loader from Manning when we started in 1986 because it had many uses and included a rock bucket, a large general-purpose bucket and a set of adjustable forks. WVCC used this machine mainly for "bull-cooking" around the jobsite, including loading and unloading trucks with the forks, loading aggregate at the batch plant, etc., and it was almost indestructible. In the early 90's, it was deemed expendable and we reluctantly sold it to Dallas Mowat at Target Products, a company that we dealt with a lot over the years for purchasing grout and related concrete products. Apparently, they needed a machine of that size to do light work around their yard, so it was perfect for them. Several years later, once we were on the Miller Creek Hydro project, we needed something just like that again to work in the main yard and our mechanic, Gerry Nelson, who was a friend of mine from North Delta, told me about a nice 966 Cat loader that was for sale in a gravel pit in Kelowna. It was owned by Andy Millar, another good friend of mine from North Delta who I went to school with. After speaking to Andy, I went over to Kelowna to look at the machine and it turned out to be the same loader that we had sold to Target! Well, it was still in beautiful operating condition and we bought it again and it served us very well for many more years. Talk about a small world!

of money we charged ourselves in company rent. This operation was a huge success, and we never once had to reschedule a pour to wait for someone's basement or driveway to be poured, which is common when dealing with local ready-mix suppliers. When we were all finished with the concrete plant, we sold it lock, stock and barrel to Greystone Ready-Mix in Langley because it was more of a stationary plant and not suited for frequent moves, which we would likely continue to require.

Another big coup at the Castle River Bridge was the supply and installation of rip rap. We got a few budgets at the time of tender but no hard quotes for this item for some strange reason. We carried a budget of $48.50/m^3, and when we started negotiating prices after our contract award, we received a quotation from Drain Bros. from Blairmore, Alberta to supply and install the rip rap for about $34.00/m^3 (as best as I can remember). This was a pure savings of over $360,000 based on the original quantity of 25,000 m^3. However, as the final design evolved, the required quantity of rip rap was reduced to 15,000 m^3, which would have been a reduction in the buydown profit alone of $145,000. Boy did that sting! After a while, I gathered up enough courage to request an increase to the unit price due to the drastic reduction of quantities and our significant loss of indirects and profit. Fortunately, both Tony Bowman and Pat Leyne at CP Rail were understanding and very accommodating of my request.

Castle River Bridge

Unfortunately, Clem suffered a mild heart attack on November 15, 1989, while working on the substructure on the Castle River Bridge. He was taken from Pincher Creek to Calgary General Hospital by ambulance where the doctors apparently told him it was very minor and resulted in no damage to the heart. I think he understood this to be just a fluke of nature and that there was nothing really wrong with his heart—that was certainly what he told us. Much to everyone's surprise, Clem was ready to go back to work a short while later. By that time, we were mobilizing to the Ashcroft River Bridge in early January 1990, and Clem became the superintendent.

KC came over from Road Link H to finish up the last remaining pier at Castle River and remained on site to supervise the erection of steelwork by Spantec and the GWIL 250-ton crane once it was all finished up at Road Link H. This was a big steel erection with the steel girders provided by Canron in Vancouver and again the erection engineering was provided by Somerset Engineering. Jim McLagan, the project manager at Canron surely lost a lot of sleep getting these very large and heavy girders delivered to the site. Once the steel was erected, KC and a composite crew from both Road Link H and Castle River formed and poured the bridge deck without incident.

In the end, we made over $1,000,000 profit on the Castle River Bridge and CP Rail was again, very pleased with our work.

Castle River Bridge After Completion

McDame Concrete Foundations - Cassiar, BC

While we were building the three bridges in Alberta, Art was up in northwestern B.C. at the Cassiar Mine building the McDame Creek Concrete Foundations for Cassiar Mining Corp. This was a complete upgrade to the existing asbestos mine in that area, which opened in 1952. Believe it or not, the entire mine closed and the town turned into a ghost town only a few years after Art finished up there. That's the mining business for you, especially when asbestos was essentially banned from most applications around that time due to serious health risks.

1989 Summary

In addition to everything else going on that year, I bought a bare quarter acre lot in a new small development in Abbotsford and had a house designed. Construction started on December 1, 1989, and I was the general contractor. Wayne Clark, Emery Baker and Earl Hearnstead, three long-term employees of WVCC, did all the framing for me while they were in-between jobs over that fall and winter. My friend, Art Findlay, who was a very good custom house builder, took over from the lockup stage to finish the house in the spring and summer of 1990. I recall one company get-together we had on one weekend where I served the beer and hamburgers and the crew helped me build my fence along with a few other exterior finishing items. I think we all went out for a nice dinner that night with the wives. To the best of my recollection, that crew consisted of Joe Mather, Pete Steiner, Glen Mosely, Blaine Benney and Korey Mather, but I may be forgetting some of them. We achieved in one weekend what would have taken months to complete on my own, so it was much appreciated, particularly because we owned a beagle at that time and needed a fence to keep her home.

Family in the Tub on Orchard Drive in Abbotsford – 1990

Left to Right – Chaylene, Me, Bill, Taralyn and Derek

1989 was quite a year and every job made a good profit, especially Castle River and Cowley Concrete, but it took a tremendous effort on everybody's part. My entire year was literally shuttling between our Langley office and the Vancouver airport for flights to Calgary, then driving back and forth between Red Deer and Pincher Creek. I was dealing with the many details of construction including managing many different major subcontracts; managing three large structural steel supply contracts as well as three separate erection procedures; sourcing enough skilled labour, equipment and materials; negotiating with owners; change orders; and resolving all payment items and claims from subcontractors and suppliers, including past projects that were not yet settled. In addition to all that, I had at least one side trip to Terrace for the gas line, one visit to the Sechelt Water Treatment Plant and various site visits for the Ashcroft Bridge and other tenders we were bidding. Those days were in the infancy stage of cell phones, so when I was driving, I was constantly dropping calls in every coulee I dropped into. Although it was far from efficient communication, I have to admit it was far better than the old radio telephones. In those days, I started using a Dictaphone when travelling so I could make notes as I drove back and forth, then I transcribed it to paper notes when I was on the plane or whenever I got the chance. I was so busy that year that when we were awarded the McDame Creek Concrete Foundations project near Dease Lake, I couldn't possibly do it, so we delegated Vern Dancy to sponsor the work with Art Penner and I never even got to that site.

Ashcroft Bridge - Ashcroft, BC

1990, our fifth year in business, was another extremely busy year that got off to a fast start with the award of the Ashcroft Bridge over the Thompson River in Ashcroft, B.C. on January 3, 1990, for a value of $5.5 million. Buckland & Taylor was the engineer of this attractive, slender bridge design. Clem ran that job with Joe Mather as his foreman and Dan Cave as assistant superintendent once he was all finished with the Castle River Bridge and the Red Deer Bridge. Bernie Buettner, Roy's son who was attending university for civil engineering, also joined his uncle Clem in May to work for the summer. Less than three weeks after the Ashcroft

Bridge's award, on January 22 we had another preconstruction meeting for the Wathl Creek Bridge in Kitimat, B.C., which was valued at $1.5 million. This was another project for M.O.T.H. with Art Penner and Paul Schlauwitz going up to build that job.

The Ashcroft Bridge had a total of four fairly tall, identically shaped slender piers, two on shore and two in the river, as well as two medium-sized concrete abutments. The river piers required rock berms and small work bridges installed for access and a combination of sheet pile and timber crib cofferdams with tremie seals were utilized to dewater the foundations. H-piles and sheet piles were installed by Quadra Construction and the river piers were constructed in the winter when river flows were at their lowest; however, this is not to say they were easy to dewater. Boulders and other subsurface obstructions like old timber cofferdams made it difficult, but eventually, after a lot of struggles, they were successfully dewatered and we were able to conventionally form the concrete piers. Pete Steiner, Joe Gardiner, Ken White and Mike Dorris worked with Joe Mather on this substructure using a local concrete supplier, Norgaard Ready-Mix. Rod Schoof was the M.O.T.H. project manager from Kamloops and Phil Munn was the project supervisor on site. Sanders Contracting, a local contractor, was subcontracted to perform the excavation, build the access roads and complete the bridge end fills. Lloyd McHarg and our 60-ton crane were mobilized from Pincher Creek to assist with the pier construction. The concrete piers were constructed through the spring and preparation for the steel erection was underway concurrently.

KWH Constructors (Peter Saunderson, Jeff Mullins and Bob Hawk) was given a target price subcontract to perform the steel erection engineering and erect the steel, working closely with WVCC on a profit and risk-sharing basis. The structural steel was fabricated by Victoria Machinery Depot (VMD) in Victoria. The girders were very wide trapezoidal tub sections that were launched into position by adding a light structural steel nose and a false bent and installing rollers on the piers. The steel sections were then jacked across the span by using a very innovative hydraulic jacking system on the bottom flange that was designed by Somerset Engineering. It took months of details and meetings between WVCC, KWH, M.O.T.H., VMD and Buckland & Taylor in order to achieve this tricky but ingenious erection procedure, as well as the necessary changes to the structural steel to allow launching. Jeff, Peter, Bob and Arvid Christianson, KWH's foreman, were a pleasure to work with, and we partnered on many successful projects in the future. The steel was successfully erected by August, and Clem and his foreman, Gord Stager, were then able to start the deck formwork.

The deck overhangs were huge on this job and required a lot of innovation, including stabilizing the top flanges of the steel bathtub girders to avoid overstressing and rolling. We had worked with Dete Mordhorst (son of Hans Mordhorst – Hansa Construction) the owner of All-Span Engineering and Jerry Weiler, another well-known structural engineer, over the last while and they were instrumental in managing the formwork design to eventually satisfy the strenuous concerns of both Buckland & Taylor and M.O.T.H. Both Rod Schoof and Jorge Torrejon from Buckland & Taylor were convinced and adamant that the deck formwork would fail due to excessive shear stresses in some of the timber members. We were all well aware of these high forces, but it was well known to us that formwork does not typically fail in shear. However, in the end, thanks to Dete and Jerry, we had to point out a little-used provision in the Timber Design Manual that allowed a designer to double the allowable shear capacity of dimensional lumber if it is visually inspected for cracks and checks beforehand, which we were more than happy to do. This little argument delayed the deck progress by at least one month, but we were eventually paid by M.O.T.H. for all the costs related to this delay as our design proved to meet all required engineering standards. Roy Buettner was particularly supportive of our position regarding this disputed deck design, perhaps because of his substantial experience working as a contractor. In the end, it all worked out just fine.

For one of the first times we knew of, M.O.T.H. specified in the bridge contract that a safety net be installed before erecting the steelwork or constructing the deck; therefore, a complete net spanning end to end of the bridge and 10' outside of the deck dimensions was installed before any of the work proceeded. In addition to the net, we installed handrails along both edges of the deck formwork as it was constructed, as is usual practice for fall protection. Once the formwork was completed and the deck concrete and parapets placed, we installed WVCC's brand-new heavy-duty rolling overhang scaffold, which was a big improvement over the last couple of versions I had used in the past, in order to strip all of the deck formwork. Before the scaffold could be utilized, we installed cable tight lines along both edges of the bridge to which every worker was required to connect their safety harness when working anywhere near the overhang. This was triple fall protection and 100% WCB approved. Notwithstanding these huge precautionary undertakings, we had one unfortunate local fellow, Philip Ghostkeeper, who was working on the overhang during the formwork stripping operation when he mistakenly thought his belt was clipped on to the safety line and attempted to lean onto his harness (which should never be done), causing him to fall down to the net, bounce over the edge of the net and land on the ground below. He was not killed, but he suffered serious injuries that lasted quite a while. The deck, along with the remaining earthworks, was finally finished in the fall of 1990—a very successful project other than that one very unfortunate mishap.

Once the new bridge opened, KWH began the demolition of the existing steel truss bridge, which was a very interesting project in itself. That subcontract was also performed on a target price basis, so there was a share of risk and reward; therefore, we were both highly incentivised to work closely together in a cooperative manner. That bridge was a six-span riveted steel truss bridge built in about 1930. It was a major undertaking to remove this bridge considering the environmental considerations with the Thompson River below.[23] The first thing KWH had to do was relocate the safety net from the new bridge to the existing bridge to meet the contractual safety requirements. Next, they removed all bridge barriers, deck surfacing, walkways, pipe crossings, handrailing and floor beams to lighten the load and provide access to the main steel chord sections. Next, they calculated the remaining dead load compression forces in the centre of the top chord of the span and jacked a load into the span using hydraulic jacks to reduce that compression load to zero. Once that compression

23 Vern Dancy, the estimator for this bid, wrote in the estimate, possibly jokingly, that the method to dismantle the bridge was to hook up a D8 Cat to one end and pull it down off the piers and into the river, then dismantle one piece at a time! As crazy as that sounds in this sensitive fisheries environment, I see videos of bridges demolished just in that fashion all the time in other parts of the world!

load was eliminated, they were able to cut the top chord member without sudden settlement and then remove sections one by one back from the centre using cutting torches and a crane from each end of the bridge. Once the steelwork was removed, we subcontracted Sanders Contracting to demolish the old concrete piers. This was accomplished by hydraulic hoe-mounted hammers basically dropping the large piers by chipping out a huge wedge at the base and felling each one like a tree. Once the piers were down, they were completely demolished and trucked away to a disposal site.

Ashcroft Bridge

While the Ashcroft Bridge substructure was being built, KC was finishing up the final details for the Road Link H Bridge and the Castle River Bridge. Dan Cave was back up in Red Deer cleaning up final deficiencies on the Red Deer River Bridge; however, we weren't done in Alberta quite yet as CP Rail was giving us extra work on both bridges including suppling, hauling and placing railway ballast on Castle River, hauling railway ties, installing safety lines on the exterior girders of both bridges and installing additional rip rap.

Wathl Creek Bridge - Kitimat, BC

Meanwhile, Art was up north in Kitimat, busy building the Wathl Creek Bridge, which was a $1.5 million contract with M.O.T.H. Wathl was a pretty simple small bridge, but Art and Paul Schlauwitz hammered the budgets and did so well that those man-hour factors could never be used again for estimating jobs, they were so good. Ken White and his brother John also worked on that bridge, along with some local workers. The bridge was completed in June 1990. I remember visiting the job during the deck pour, and it was a perfectly smooth operation. Towards the end of that job, Paul Schlauwitz was getting married and required two weeks off, so Pete Steiner came up to help wrap things up with Art and Bernie Buettner, mainly finishing the parapets and installing the railing. George Lomas was the project supervisor for M.O.T.H., and I know that Art and George got along very well throughout this job. If only all of our jobs were this straightforward and went this well! By this time, Art was well on his way to becoming an excellent superintendent.

Snip Gold Mine - Eskay Creek, BC

Before Art had even completed at Wathl, we were awarded another mining job from Cominco at Snip Gold, which was located way up in northeast B.C. on the Iskut River. This was a very remote, fly-in site with a small camp to live in, and it looked like Art was our man again. I think Bernie and Paul stayed behind and cleaned up Wathl Creek, including removing and dismantling the existing Bailey Bridge. On the Snip Project, there were three separate contracts— the first one for mill foundations, the second for yard piping and the third for the service complex, so there was a mix of civil division people and some mechanical division trades. The total value of the work completed at Snip was $2.0 million. This was yet another job we did for Merit Consultants, who was a great client to us over the years and Robin Calder was in charge of this project for Merit. The concrete structures in the original contract included a crusher building, concentrator building foundations and a ball and sag mill. Some of the workers Art had on this project were Joe Mather, Joe Gardiner, Emeric Domokos and Blaine Benny.

Once the initial concrete work was completed, I know that Art didn't have a lot of fun on this job with the mechanical trades, but it didn't last for long and he was off to greener pastures soon enough. While that job was underway, I flew to the site at least once, landing on that little scary gravel airstrip on the banks of the narrow canyon created by the Iskut River. The little strip seemed to aim right at the rocky face of a high mountain, which didn't give me any comfort if they had to abort a landing. Also, there were literally parts of crashed airplanes alongside the runway, which added to my anxiety! I remember another time I was flying in to Snip from Vancouver airport on a small private chartered plane from Central Mountain Air when we couldn't land due to poor visibility and the pilot flew all the way back to Vancouver! What a complete waste of a day that was. The other interesting thing about this gold mine was that there was no road to it, although they had a hovercraft that travelled upriver to the site and back down to the ocean at Wrangell, Alaska. The hovercraft was capable of transporting freight, personnel and eventually gold concentrate. Otherwise, Hercules aircraft could be scheduled to bring in heavy equipment and supplies, but that was very expensive. Snip Mine was a very successful hard rock underground gold mine that took out a total of 1.1 million ounces with average concentrations of gold of .88 oz/ton over its operating life of nine years. (That translates to $2.8 billion CAD at today's gold price and the present US conversion!) For comparison, the Premier Gold Mine was never really very successful, and they were actually talking about shutting the entire operation down before we finished constructing the mill site![24]

Pine Pass Bridges - Chetwynd, BC

By late July 1990, we were awarded another project from M.O.T.H. with a value of $3.4 million for the construction of three new bridges in the

---

24 At the time of writing this book, the Premier Gold mine facility is being retrofitted with new mills to process material from local mines, so it isn't dead yet!

Pine Pass, located just south of Chetwynd, B.C. We were already pretty booked up demobilizing Castle River, Cowley Concrete and Wathl Creek; completing final concrete finishing at Red Deer and into full construction on the Ashcroft Bridge and Snip Gold Mine; and building a concrete retaining wall for the Surrey RCMP, so we were running out of our regular staff. As a result, we hired a new man, Jack Nelson, to run this job along with Richard Evans, who had just completed his work in Pincher Creek with KC Shenton. Paul Schlauwitz joined Jack as carpenter foreman along with Ralph and Mike Schlauwitz who all lived in the Dawson Creek area, which wasn't too far away. We purchased a brand-new portable Fastway batch plant and cement silo for this project from Ideal Manufacturing in Billings, Montana, and we set this up near the bridge site with Carlos Madiera running the concrete operation with our mixer trucks and loader from Cowley.

Bill Eisbrenner was the project manager in Prince George and the site inspectors were Gabor Veres and Ted Peters, who I had worked with in Penticton when I was with M.O.T.H. All of these guys were very professional and very much solution-oriented, which was a good thing because we had some problems to resolve on this contract. There were two three-span steel girder bridges and one simple span concrete box girder bridge in this contract. The piers for the two longer bridges were designed with extended churn drilled 30" diameter pipe piles, similar to the Clearwater Bridge design that we pioneered with Franki Canada a number of years prior. Piles were driven by MCL Piledriving (Dave Mitchell), and overall, these bridges were fairly simple structures and the substructures were completed by late fall. Other than a significant layout error while installing the pipe piles, the substructure went smoothly. This survey error involved a "new" method of survey and layout at that time using solely coordinates rather than measuring distances and angles. The theory of this type of survey is perfectly sound, and it is extremely common to this day with modern survey equipment; however, it appeared that, in that instance, there was no backup survey check or redundant management scrutiny to confirm the correct layout before piledriving began, and this resulted in an embarrassing and costly pile removal and relocation operation. (Remember the extensive redundancy checks required of me at the Taghum Bridge back in

Nelson?) If my memory is correct, Jack Nelson, while a good guy to deal with, moved on after the completion of the substructure.

The concrete girders were supplied by Western Concrete Products in Armstrong and the structural steel was supplied by Rapid-Span from Vernon. Dave Giles, who was an experienced steel erector from the Smithers area and formerly with Gisborne Industrial, provided us with a subcontract price to erect both the concrete box girders on Link Creek as well as the structural steel on Solitude #1 and #2. We used Lloyd McHarg and our 60-ton Koehring crane for this erection and that work was undertaken off of berms over the winter when the water was low. Art came down once the civil work at Snip Gold was completed and supervised the completion of the steelwork erection before he got started on the deck construction with Paul Schlauwitz. We ended up renting Kingston Construction's deck machine and Clem came up, set it up and ran it for us for the deck pours with Emeric Domokos and Jose Costa, our main finishers. Gordon Harvey, the senior deck man from the Bridge Branch, was also on site before and during the deck pours and assisted us to get everything set up, including adjusting the concrete mix designs, to make it all run smoothly. Everything went very well on all of the deck pours and Art had the entire project all wrapped up successfully by the end of July.

Mt. Hundere Mine - Watson Lake, Yukon

Late in the day on August 30, 1990, which was a Thursday night before the long weekend, I received a telephone call from Mike Crossey at Kilborn Engineering, who enquired whether WVCC was still interested in bidding work on the Mt. Hundere Mine, which was going to be out for tender very soon. As a normal part of our business development planning, Paul Manning had made some enquiries months before when we'd heard about this new potential mining project. I told Mike that as far as I knew, we were still very interested but Paul had left for the day, so I would get Paul to call him back in the morning to confirm. For background, the Mt. Hundere Mine was a proposed lead-zinc mine located at the 4000' elevation 45 km north of Watson Lake, Yukon. It was the owner's intention to commence construction immediately and have the concrete work underway over the winter and completed by June 30! I immediately called Paul and told him about my call and then expressly told Paul that we were already way over our heads and we could not possibly take on any additional work that year, "no ifs, ands or buts"! We already had 10 jobs awarded that year after nine the previous year, with many of those jobs running concurrently. He understood my concern, which was not only due to the fact I was overloaded, but also because we had run out of available, fully experienced superintendents, let alone our regular tradespeople and equipment. Nevertheless, we eventually agreed we would have to submit a "courtesy bid" as we could not really turn away a request to provide a proposal to a potential mining client, but he assured me we would not be low bid on the job, so not to worry. Several weeks later, shortly after submitting our proposal, we received another call from the owner, Curragh Resources. They wanted to meet with us along with Kilborn Engineering, the project manager, right away. Paul and I went down to Kilborn's Vancouver office and did our best to hold our position, clarifying what was included in our bid and more importantly, what we had not included in our pricing. We provided a lot of clarifications in our proposal, something we were becoming very adept at with these mining projects because so much information was not yet known. Well, to make a long story short, our clarifications were not only accepted, they were very much appreciated. We were awarded the contract in the amount of $4,263,000 after that meeting, and we had to start mobilizing in late October! Now we were under extreme pressure to perform.

Without any of our senior superintendents to run this job, we decided to name Dan Cave as superintendent, as he had been working mostly as an assistant superintendent for us over the past two years. We volunteered Vern Dancy, our civil estimator, to be on site as a project manager for at least the start-up of this job (we figured we couldn't bid much more at that point anyways). Unlike the McDame Project, I was still able to sponsor that project even though I remained engaged with the Ashcroft Bridge, the Pine Pass Bridges and cleaning up all of the Alberta work. Bernie Buettner and Mike Minshall, a new hire from BCIT, were also on-site assisting Vern and Dan with survey, logistics, purchasing, etc. Along with Vern, Dan, Bernie and Mike we were able to get Pete Steiner and Paul Schlauwitz as foremen at the beginning of the work and Joe Mather came up a bit later to run the crusher and mill foundations once he finished up the substructure in Ashcroft. Paul was in charge of building the large coarse ore bin, which was a square concrete structure about 80' high that had to be entirely hoarded in to be able to complete it at those temperatures. Other crew members who we managed to get on that job included Wayne Clark, Nick Maskery, Emery Baker, Mike and Ralph Schlauwitz, Calvin Fauth (an ex-Goodbrand carpenter), Gord Lakeman (who we found in Whitecourt), Corey Mather, John Jaconen, Ken White, Hardy Hibbing, Guy Dobbin, Mark Stroeder, Gus Gudmundson, Emeric Domokos, Blaine Benney, Carlos Madiera and Mark Wilson, whenever we could shake them loose from the other ongoing work that we had on both the Pine Pass Bridges and the Ashcroft Bridge. Concrete was supplied by our crew with another rented batch plant located on site and all of the crew lived in a small camp located on the mine site. On that project, we worked closely with Kilborn's staff including the project manager, John Wigle (again), along with Alain Catteau and Graham Clow, who were working on site.

The fact that this new mine infrastructure was built over the winter at this elevation, that far north in such a remote location and in such a compressed schedule with our so-called "B-Team" was quite an achievement, especially since this job ended up making over $1,000,000 in job profit. So, it was quite a wild end to 1990! That was the eleventh project that year and our most profitable. That year, our company undertook almost $17 million worth of work and we bonused out $1,000,000 to staff and key people!

After five years of operation at WVCC, we had completed $49 million worth of work and also assembled a great, well-trained staff to manage the work, as well as a very diverse, competent and dedicated crew of skilled tradespeople. We had furthered our already considerable experience in both highway and railway bridges and in industrial work like the pulp mill and sewage and water treatment plants and we gained a significantly strong reputation in the lucrative mining industry. But it was sure a lot of hard work and considerable stress. Around this time, Marny Woolf (daughter of Karen Woolf), who started with the company in January 1990 in the accounts receivable, subcontract accounts payable and costing department, continued to show tremendous potential in our business and we decided to transfer her over to the construction division to be my assistant. This strategy was intended to reduce the natural bottleneck in workflow and capacity that was created at my position, and it was a great move. Marny was a very capable, reliable and hard-working assistant to me, and she had all of the fantastic qualities of her mother, Karen, including perseverance and accuracy.

One other accomplishment we made in this era was the establishment of a corporate Policy and Procedure Manual. Paul, Bill, Marny and I worked together to produce this extensive binder that covered everything from A to Z so that the operation of the various functions within the company were transparent and known to all staff. It was intended to provide consistency to everyone in the company as well as a clear guideline on how we expected things to be done, whether it was purchasing policies, methods for estimating, costing, equipment maintenance, writing daily reports, how to handle extra work, and on and on. It was a comprehensive undertaking and we were all quite proud of producing such an all-encompassing policy manual for a company of our small size.

In those early years, although we were operating with quite a small staff and everyone was extremely busy, we always tried to make time to provide training and learning experiences to all of our staff, even those in head office administrative duties. We arranged field trips and seminars for team building between all of the staff and many of them were also very educational. These experiences often revolved around our annual general meeting and included tours of an old operating foundry, a bolt manufacturing facility, a welded wire mesh plant, a precast concrete plant, a brick-making plant, etc. We also had in-house training sessions where we invited special speakers including Paul Ridilla, a travelling construction guru from the United States, and even a psychologist from Royal Columbian Hospital. Our own in-house training sessions often included discussions on sections of the Policy and Procedure Manual, such as job costing, jobsite efficiencies, concrete mixer truck maintenance and preventative maintenance on a variety of other equipment. These were always interesting and I think well received with positive benefits.

CHAPTER 18—
Trip to Europe

IN THE SUMMER OF 1991, just as we were winding down the Ashcroft Bridge, Pine Pass Bridges and Mt. Hundere, I decided I needed a break, and I took a full six weeks off to travel to Europe with my young family.[25] Using my Aeroplan points that had accumulated, I booked four business class seats on Air Canada from Vancouver to Athens, Greece with one stopover in London, England. Being the fairly adventurous type, I decided to rent a motorhome in London for the first three weeks so we had the ultimate in flexibility and we wouldn't have to book hotels every night. Well, it turned out that motorhomes were quite rare in England in 1991, so it became quite the experience! While picking up the motorhome close to Heathrow Airport, the staff spent an hour or two going over the various procedures of the motorhome, such as the operation of the fresh water and toilet holding tanks, the propane heater, the stove, oven, etc. so I was thoroughly comfortable with those functions. However, once I left the rental facility and got onto the road, I realized they never once gave me any advice or directions whatsoever on shifting a five-speed transmission with my left hand with a gutless 2000 cc engine or driving in roundabouts! Of course, I knew that I had to drive on the left side of the road, but I

25 Annette and I travelled with Derek and Taralyn, but we left Chaylene home with her grandparents because she was only three years old. While I have always regretted the fact that she never got to go to Europe with us, we made it up to her with subsequent trips to Hawaii, Mexico and all over B.C. Later in life, she became a seasoned traveller with her husband Shannon, visiting literally many dozens of countries all around the world.

had no idea how to operate in the roundabouts, and this was both scary and dangerous, especially at the speed the locals drove. After driving aimlessly for about one hour and not really getting very far, I pulled into the parking lot of a pub that I found, went inside and walked up to the bar (no, I did not order a drink to settle down!). The bartender took one look at me and asked me what was wrong (I must have looked white and in shock). I explained I was from Canada and had no idea how to operate in the roundabouts. After a good chuckle, he took me aside to the end of the bar and explained it this way using his fingers on the bar to illustrate: "As you approach the roundabout you look right and merge into the circle—fuck the left!" Ok, I think I got that. When I got back into the motorhome, I started driving again and as I approached the first traffic circle, I repeated the mantra out loud, "Look right and fuck the left!", even though I was with my wife and kids! This worked beautifully and for anyone who has been in that country, and more specifically, in that area north of Heathrow where there are roundabouts at least every ¼ mile and they often actually run into each other in succession, you need to get this practice down and get it down quickly or you will be hopelessly lost.

Now, the next big learning curves were map reading and being able to navigate the roundabouts so that I took the correct exits to our planned route. You see, in Canada, when you leave the city of Vancouver, you start seeing signs immediately to Hope, B.C. and you can easily follow those directions. However, when you leave London, you do NOT see any signs to Manchester or Glasgow, but you do see signs to Sipson, Northolt, Luton, Iver Heath, Uxbridge, Eton College, Hayes, and many more towns and villages in the immediate area that you have likely never heard of! Therefore, you need to know in advance what little towns are along your planned route. This requires a pretty extensive degree of map reading, something that definitely cannot be safely done by the driver while grinding through five gears with your left hand and driving on the left side of the road. So, as the day unfolded, even though we stopped and checked the overall route we wanted to take, it turned out Annette could not read the jam-packed map well enough to tell me in advance what the required exits were and that added an immense amount of additional stress. Let me add that the road atlas I purchased in advance for Britain was not useful for navigating anywhere around London, so I also bought a road map book of London, which turned out to be very difficult to read because it basically showed a smear of curving black lines everywhere! I ended up promoting my son, Derek, to chief navigator in the front seat and that worked out a bit better, but it was still a tough job.

After a few hours of stressful and tiring driving, we arrived at the small town of Dunstable, which was 36 miles north of Heathrow on the A5 Dual Carriageway, where we needed to stop for the day out of pure exhaustion and jet lag. I had a travel book with me to look up camping facilities in that town and found one not far away, so we decided to have dinner in a local restaurant near where I could easily park the motorhome. That was the most tasteless dinner I had ever eaten in my life. I had heard that English food was a bit bland, but this was ridiculous. Maybe it was because we were so tired and stressed with our new experience. When we were done with dinner, I drove the short distance over to the campsite where we discovered they had an ancient stone archway in the entrance to the camp and the motorhome would not fit through it! (Do I look like Chevy Chase??) After speaking to the camp operator, he told me we could park down the road in the back of the White Swan Pub for a quid, so we headed there promptly. Once we arrived, I went in and told the bartender of our dilemma, and he said it was okay to park there for the night. It didn't take even 15 minutes to have all the beds set up, our teeth brushed and the doors locked, and we were instantly asleep. That was 8:00 PM local time, and no one stirred until 12:00 noon the next day.

After sixteen hours of uninterrupted sleep, I felt pretty well-rested, but overnight I'd had an epiphany that we shouldn't be driving this motorhome in England and that this whole idea was a big, dangerous mistake. I calmly decided I would return the motorhome that day, and we would be more conventional by just renting a car and staying in hotels and B&Bs. After finishing breakfast, we all got in the motorhome and headed back to Heathrow Airport, but this time driving was much easier as I was getting used to the stick shift and roundabouts and Derek was getting better at navigating. When we got back, I tore a strip off the manager for letting me out of the facility without any training on roundabouts or any explanation about the basics of driving in England, and I told him that I wanted a full refund. Well, to make a long story short, he said the only way we would get a refund would be if he was able to rent the motorhome to someone else, and based on the short notice, that wasn't very likely. Based on that response, I told him I would just keep it but not to expect to see it in one piece when I returned. I was pretty sure we would have a crash of some sort, but I was fully insured up to the gills. We were forced to head north again and try to get out of London one more time. The day progressed and the further away we got from the winding mess of narrow roads around London, the easier it became and the more comfortable I got. By the end of that day, I felt like Parnelli Jones, and we made it all the way to Nottingham in central England where we found a beautiful camping spot right along a stream. We were trying to take intermediate roads and definitely avoid the big cities, but the roads in England were nothing like any in North America. One small issue I had on this leg of the trip was when I got momentarily confused about which lane on the motorway was the slow lane. The fast drivers in the Mercedes and BMWs certainly got me straightened out very quickly with excessive use of their horns and shaking fists! Boy they were upset! However, from that point on, the motorhome was no problem at all. It was great that we didn't have to haul suitcases around every night wherever we stopped and were able to cook a lot of our own meals.

As we headed north, we went through Sheffield with all of its old steel mills, through Leeds and then eventually stayed in the ancient town of York where we visited the fantastic York Minster, a cathedral built in 637 AD. Then we headed east out to the sea and to colourful and quaint places like Scarborough and Whitby. Continuing north through Newcastle, we made it to the beautiful and historic city of Edinburgh, Scotland, which was one of the highlights of the entire trip. Up to that point, the weather had been warm and sunny, which made travel very scenic and enjoyable. After touring the city for a full day and enjoying the sights, architecture and culture of Edinburgh, all while our laundry was being done, we headed north over the Firth of Forth Bridge and up to St. Andrews, where I recall we were able to book a hotel room for a nice change.

The next morning, I went down to the Old Course, the birthplace of golf, rented some clubs and played a round with an American father and son from Ohio.[26] I recall the cost for the green fee was £30, which was about $60.00 CAD at that time, and that was one of the most memorable experiences in my entire life. Having played pretty well, I finished the Road Hole (17th) and was walking over the Swilcan Bridge up the 18th fairway where I could almost hear the cheers and applause from the spectators lining the right side of the fairway as if this was the final hole of the British Open. As I walked up the fairway to make my second shot, I scanned the imaginary crowds to see if my family was waiting to see me play this most famous of all finishing holes in all of golf, but they were nowhere in sight. After finishing the round and saying goodbye to my new American friends, I spotted my kids way over to the left, playing on the beach and in the sand dunes, completely unaware of the significance of this moment. Having a beer after the round at a nearby pub was like icing on the cake! I loved every bit of St. Andrews because it had so much British history and it had hosted the British Open many times, which remains one of my favourite tournaments to watch every year. Now I was part of that history, at least in my mind.

We left St. Andrews and headed west, crossing the Kincardine Bridge over the Firth of Forth and into Glasgow. We found Glasgow to be a very large and crowded city that was far from as beautiful as Edinburgh, so we continued driving past Port Glasgow, Greenock, Troon and Prestwick, then down over to Dumfries, Preston and on to Liverpool. We enjoyed a nice formal lunch at the Marine Hotel, which was located right on the famous Troon Golf Course, and I could not believe so many golfers were playing in the very windy conditions. It was so windy that the flagstick was literally bent over horizontally! The highlight of our Liverpool tour was the Beatles Museum, which was very good. We then headed down to Bristol, which is a great-looking seaside town, except for the fact that it was

26 It wasn't quite that easy. After rounding up my rental clubs and paying at the pro shop, I reported to the starter in the little starter's shack near the 1st hole and let him know I was ready to play. After about 45 minutes, and a few twosomes and threesomes teeing off on their own, I went back to see this grizzled old Scottish starter and enquired why I hadn't joined one of those groups. He gruffly informed me that it wasn't his job to set up a foursome for me, that was my job! Then I asked to join the gentlemen that were heading to the tee and I was off.

cool and windy, not at all like the summers we were used to. The next real highlight of the trip was our visit to Bath, which is an ancient Roman city that has hot springs and Roman baths throughout the old city. However, unlike the fantastic weather we enjoyed in northern England and Scotland, it was pouring rain down there, and we decided to head out of town to find a place to camp. On the way to Salisbury, we smelled smoke and found the motorhome was quite smoky in the back. I pulled over instantly (as best I could do on a narrow road with no shoulders) and went back to inspect everything. I found a rear wheel axle bearing had heated up and was literally on fire. I cannot recall if I used the fire extinguisher or a bucket of water, but we eventually got the fire put out and I called roadside assistance to tow the crippled behemoth somewhere to get repaired. As the smoke had diminished, Annette and the kids waited in the motorhome with the windows open while I hitched a ride back to Bath in the pouring rain! Once I got to the town, the driver let me off at a local pub where I made a call to AAA who said they would dispatch a suitable tow truck for the motorhome and pick me up at the pub on the way. Unfortunately, to my complete dismay, it was 6:00 PM and evening closing time at the pub (British social engineering so that the patrons would get home for dinner with their families), and I was forced to wait outside in the pouring rain as most old buildings in England do not have any eaves or awnings to stand under! Grrrrr! Eventually, a tow truck driver showed up. He looked like some self-imagined rock star with hair like Keith Moon, no shirt and work coveralls undone to his waist. We drove to the motorhome, and he seemed very interested in Annette with his sly, toothy grin. We got the motorhome loaded and all four of us hopped into the cab with this limey greaseball who drove like a madman into town where there was a repair shop. Of course, it was now closed for the day. After I changed my wet clothes, we gathered our luggage, checked into a local hotel that was close by and went out for dinner.

Once we got the motorhome back on the road the next day, we headed to Stonehenge, which was incredible to see in person and everyone loved it. After that, we headed towards London where we spent a couple of days touring around. As I recall, we parked the motorhome well outside of the city, took the train in each day and did a lot of sightseeing on double-decker bus tours and even a horse and buggy tour around Windsor Castle. We walked miles through the city and took photos at every corner. There is a lot to see in London, and we saw a lot, including the British Parliament, Big Ben, Tower Bridge, The Eye, Buckingham Palace, Hyde Park, Westminster Abbey, Windsor Castle and a few pubs that we were able to enjoy. The trains in and out of the city were fantastic, and we were able to move around the city with ease, although we did our best to avoid rush hours.

After the tour of London was done, we headed in the motorhome to Dover to take the ferry over to Calais, France and made a beeline to Paris. At least over on the mainland people were driving on the correct side of the road, but now all signage was in French and people were driving even faster than in Britain! This trip was becoming very much like the movie European Vacation, especially when we pulled into a small town for lunch and ordered what we thought were hamburgers and received something like the worst chuck steak you could imagine. I think the waiters thought we were Americans and treated us accordingly. Even though I had studied French in high school for four years, I could really only communicate simple nouns. I could generally get by in a little grocery store, but it was much more difficult understanding French menus and ordering meals from people who apparently didn't know (or pretended not to know) any English whatsoever. "Fermez la porte" or "Ouvrez la fenêtre" never actually came into use, so I was pretty well hooped.

When we got near Paris, we found a campsite in the town of St. Germaine, which we made our home base, and on a couple of days, we took the train into the city to see the many sights. Paris was a fantastic city that was easy to get around in. We went up to the top of the Eiffel Tower and spent almost a whole day in the Louvre Museum where we saw the Mona Lisa and many other famous paintings and sculptures. We also took at least one bus tour to get an overall view of the entire city. One thing we quickly learned on the mainland was that it was very expensive to sit on the outer ring of seats in bars and restaurants near the busy traffic circles, as the prices could be 10 times more expensive than inside the restaurant. Don't laugh until you have paid the equivalent of $20.00 USD per Coke for two kids and even more for a draft beer in one of those smoke-filled cafés! One afternoon, after another full day of touring Paris, we stopped at a small café near the campsite where the motorhome was parked and ordered some drinks. The kids thought it would be fun to have a duelling match on the patio with the baguettes that we had bought earlier for dinner and the locals thought we were crazy, but the kids loved it.

Once we left Paris, we headed south for the French Riviera, travelling through the beautiful, hilly agricultural countryside of Orleans, Lyon and Cannes, then on to Antibes, where I anticipated perfect beaches with warm water. We located a semi-decent campsite right across from the beach, but the beach area was far from what I had expected. The beaches were all pebbles with no sand and it was quite similar to Mexico in regard to sanitation and cleanliness. The washroom at the beach consisted of a small plaster building with no doors, a hole in the centre of floor and a short hose nearby. No privacy dividers and no toilet tissue! On the other hand, the beaches were topless, so as long as you didn't have to visit the washroom, a red-blooded Canadian boy could enjoy the beach to some degree. The temperature of the water was somewhat disappointing as well, but it was only the first week in July, so being healthy, rugged Canadians, we did enjoy swimming a bit.

The next day, we headed west along the rugged coastline, which was very scenic and beautiful. We passed by St.-Tropez and eventually got through the big port city of Marseille and a little further along, we found another nice-looking beach resort at Sète. This campsite was much nicer and the place had a great swimming pool (also topless), which was all good because the sea was freezing cold for some strange reason. Continuing west, the next day we travelled through Perpignan, crossed the border into Spain and crossed through Barcelona, then turned inland and headed across the Iberian Peninsula towards the Bay of Biscay in the Atlantic. As we travelled through Barcelona, we found massive construction everywhere with new buildings, subways, highways and interchanges, so to avoid the heavy traffic, unfortunately, we hurried through town without seeing that much. On that route, we travelled through Zaragoza, Pamplona (where the bulls run) and then down the big steep hill to San Sebastian, on the sea. Driving west along the Atlantic past Bilbao, we found a great campsite right on the beach at Santander. After enjoying one night on the beach, we turned around and headed back through San Sebastian, which was a beautiful ancient village with a spectacular blend of architecture right on the water. We then crossed the border back into France and headed north. The drive through western France was nice; most of it looked like plantations of forests and the remainder was hilly agriculture and vineyards, like so much of France. On that route, we travelled through Nantes, Rennes and then for some unknown reason (remember, my nickname is Backwoods), I turned a bit northwest on more secondary roads, following my instincts, but not sure what I was looking for. After some time driving in this flatter, low-lying countryside, I spotted something in the far distance and then continued on that path until we could eventually see the castle on the island at Mont St. Michel. The closer we got, the higher it rose from the mudflats of the ocean and the more mysterious it became. I had either seen pictures or war movies of this castle before and wasn't sure what it was called, but it was really quite remarkable to see it in person. Unfortunately, the tide was fairly high at that time, so you couldn't walk all the way to the island, but it was spectacular to see up close. From Mont St. Michel, we headed to Le Havre for the ferry back to England where we had to return the motorhome in a couple of days.

The ferry crossing the English Channel was again uneventful, but we could certainly feel the waves and tide moving the boat as we plowed our way across over 120 miles of open seas. Once we got back to Heathrow, we sorted all of our things, said goodbye to the motorhome and checked into a hotel for the night. All in all, the motorhome was a great idea and the four of us enjoyed it immensely. It was very convenient for cooking when we felt like it, and we could always find a spot to camp without too much trouble, except for that first night when we were so jet-lagged. After getting settled into a hotel, I went off on my own by train to a nearby town called Hounslow to try to purchase another large bag that we could use to store a bunch of things in the lockers at the Heathrow Airport that we would not need for the next leg of our trip. Well, Hounslow turned out to be an area with a heavy population of apparently unemployed black people because the streets were just as packed and littered with garbage like when I was in Harlem! It was shocking to see and certainly not what I would expect to see in England, but I managed to buy the bag and get back to the hotel unscathed. Into this bag, we deposited all of the cool weather clothing, rain jackets and boots as well as all the tour books for England, Spain and France because our next stops were Rome, Italy and points to the south where it would be nice and warm.

The flight to Rome was the only thing actually booked on the entire trip because, as I was told many years later, I like to "fly by the seat of my pants", although again, it was business class! Our best directions once we got out of the airport in Rome were to "turn left immediately out of the station and don't talk to anyone until you get several blocks away." Whoever gave me that advice was very clear that we should not turn right or I would be walking the family into a potentially dangerous situation. So, there we were, four skinny, white-skinned people with blonde hair walking closely together with our pack sacks and money belts securely strapped on and our heads down until we got safely away from the airport and all the hawkers, conmen, and swindlers. Once we were clearly out of any impending danger, I found a little café that we piled into (the inside, not the perimeter outdoor seating), dumped off the packsacks and ordered two beers and two Cokes. Once I had a gulp, I took off in a run (yes, that was a long time ago) to find a hotel nearby that would be our base for a few days. Down one or two blocks, then over a couple of blocks in the other direction, I eventually got about five quotes. I selected one hotel with a room that I think was on the fourth floor, that was reasonably priced (for Rome that is) and that looked like it had a nice view and a rooftop sitting area. I jogged back to the café and found the rest of the family choking from all the smoke, so we left promptly. Again, the bill was shocking as the four drinks added up to about $60.00 CAD! Once we got to the hotel, we unloaded our bags, went up to the rooftop lookout and relaxed for a while with an ice-cold Peroni beer. After a while, we went out for dinner, where I learned another travel lesson (this was well before Rick Steve's television series on European travel). We found a nice little pizza place with outdoor seating not too far away from the hotel, and we thought it would be nice to have a real authentic Italian pizza. The menu was, of course, in Italian, but I couldn't find shrimp and mushroom anywhere (just kidding), so we ordered some plain cheese and tomato pizza and, while different than what we were used to, it was great and everybody loved it. The kicker came when the little slick-haired waiter brought the check and included a "coperto", which seemed to be some kind of service charge, because we understood from our tour books that tipping wasn't common or certainly not expected in Italy at that time. This "coperto" added up to a considerable amount of money—way more than a tip might have been. I immediately thought this was another rip-off like the sidewalk cafés because we were tourists, so I whispered around the table that we would be leaving quickly as soon as everyone was finished and ready to go. Remember, Derek was ten years old and Taralyn was eight, so their eyes were wide open when I told them we were going to make a run for it! After counting out the correct amount of money in lira without the "coperto" and placing it under my plate, I signalled to everyone and we were off at full gallop. We ran for a few blocks, and I felt fully redeemed—I wasn't going to let these people rip me off anymore. Back at the hotel we slept like babies after another full day of travel from London to Rome.

The next day, we spent the full day walking around the city of Rome seeing many of the tourist sites such as the Coliseum, the Roman Forum and the Trevi Fountain, as well as many beautiful piazzas, ancient ruins and cathedrals. The streets were packed with tourists of every nationality imaginable and street musicians and entertainers all about. You had to be wary of gypsies who could rob you blind, and we actually saw one gypsy roughly arrested by the police for pickpocketing right near us, just adjacent to the Coliseum. As much as we got used to the motorhome and enjoyed its mobility, it was nice for a change to have a home base where our beds were made for us and we didn't have to cook, and we could also head back to the hotel for a break in the day. After a couple of days of touring around, seeing a lot of the sites of Rome and enjoying some great meals, we were on the move again, using our Euro Rail Passes and taking the train south to Naples. From there, we boarded another bus for the short and windy trip to the beautiful seaside town of Sorrento, near the Amalfi Coast. This was again, a highly recommended side trip, but it was a fairly busy tourist place with hotels and restaurants lining the narrow streets that ran along the edge of the cliff down to the Tyrrhenian Sea below. We found a nice hotel with a pool right on that cliff. It had a fantastic view of the port and rocky shoreline below, and it didn't seem too expensive. After settling into the hotel that afternoon, we headed out for an early evening walk around the town, and it appeared that everyone in town had the exact same idea. It turned out it was an Italian custom for everyone to walk around the town at this time of the evening, all dressed up in their finery with their date on their arm. There were street musicians and many people were outside of their restaurants showing off their menus and offering specials to entice early guests to come in and enjoy their hospitality. After a pretty good walk around town, we were approached by a lovely older Italian woman on a narrow, crowded street. She seemed genuinely eager to have us come to her restaurant for dinner, enticing us with all kinds of specials, and by this time, we were happy to oblige her. I think that was the night I discovered cannelloni. It was fantastically flavourful with the ricotta cheese and delicious rich tomato sauce, but it was the texture that was perfect. I think I ate cannelloni most of the remaining nights in Italy, and it is always my first choice in any Italian restaurant to this day. With some nice, authentic Italian red table wine, the dinner was perfect!

Our second day in Sorrento was very similar to the first day, although we rented a taxi and went for a drive down to the Amalfi Coast, which was very scenic, yet very busy. I am glad we didn't have the motorhome for this little adventure because the taxis could hardly make it through the traffic and around all of the tight corners. Being creatures of habit, I seem to recall we went back to the same restaurant for dinner that night, and you know what I ordered! Because we were leaving the next day, we had an early night and enjoyed our last night in that hotel with its fantastic view of the sea below, the glimmering lights of Naples and the islands in the distance across the bay. In the morning, we went down to the lobby café where we all grabbed a bagel and a juice or a cup of tea from the continental breakfast area as we had done the previous morning, assuming that this was a complimentary service included with the hotel rate. Upon checkout a little while later, I learned that these little breakfast snacks were not complimentary; in fact, they were quite expensive, priced at 27,000 Lira per person, for a total of 216,000 Lira, which amounted to $20.00 USD per person per morning! That added up to $184.00 Canadian (in1991 dollars!) for the two days and the most anyone had was a bagel or a bowl of dry cereal and a cup of tea! Again, I was furious with these expensive little Italian tourist deceptions, but what could we do but take it on the chin and know that it would be a good story to tell some day? I guess I should have learned Italian before this adventure because apparently there was a non-descript sign on the wall behind the ferns that explained in Italian the continental breakfast was not gratuito! (Colazione non inclusa)

After taking the bus back to Naples, we headed out by train to a beach resort called Vieste across the width of the country on the Adriatic Sea. This was a place that some English-speaking Italians, who I think we met at the hotel, told us was a beautiful beach area where Italians holidayed and where we could avoid the Americanos, Germans and British touristos, something that was now becoming quite appealing to us. The train took us east to the town of Foggia, where we hopped on a big, comfortable travel bus that took us out to Vieste. The town of Vieste is located on the tip of the Promontoire Gargano, which is a solid rock outcropping sticking out into the Adriatic Sea, just above the boot of Italy. The bus trip was so hilly and extremely windy and narrow that the bus had to toot its horn on every corner and it was literally sickening for the kids. It was about 90 km long and quite scenic, but that did little to help their motion sickness. However, once we arrived it was absolutely beautiful! There were large limestone monoliths rising from the beach way up into the sky and the beaches were filled with sun chairs and umbrellas with warm, inviting aquamarine water lapping on the sand. Located on the edge of the rock outcropping was the quaint and ancient little beach town of Vieste with all its small hotels, cafés and restaurants that we were able to enjoy for a couple of days. We had dinner on an ancient outdoor patio carved from stone right near the edge of the cliff above the beach, which was stunning. After enjoying a couple of days relaxing on the beach and wandering in the little town, we headed south on the train down to Brindisi, which was the main ferry terminal to our next stop, Greece!

Brindisi was a bustling little ancient seaport on the heel of Italy that was packed with tourists—mainly young backpackers travelling by commercial ferry between Italy and Greece. Our destination was Patras, which was a crossing of about 16 hours, so we would have lots of time to unwind, stretch out, read and maybe nap as best we could on the main deck. They may have had cabins for this length of trip, but we didn't bother getting one for all four of us. The trip was smooth and uneventful with lots of happy young travellers. Once we arrived in Patras, we grabbed a Mercedes taxi to take us to a resort area called Kyllini Beach, which was highly recommended by someone we'd met earlier in our travels. This was about a one-hour taxi ride from the ferry terminal. By the time we got to the beach area, we were all tired and not sure exactly what we were getting into, but the taxi driver took us to the front entrance of a very modern-looking resort located right on the sea. We enquired whether they had vacancy, and they had a room, although the staff were looking at us kind of funny. The cost was shocking (about $300 CAD, as I recall) but we needed to get a good night's sleep before making any further investigations or decisions, and this looked like a real nice seaside vacation place for our first experience in Greece, which I generally understood to be quite affordable. We noticed in the lobby there were a lot of German-speaking tourists; however, we carried on to our room and came down a while later for a late dinner. At dinner, we found ourselves completely immersed in German culture with German menus, German music playing and many blonde, tanned, muscular German men and women frolicking around the place. We quickly learned that this was a German resort and that we were probably the only English-speaking people in the entire hotel! Faulty Tower's sketch on "Don't mention ze var!" quickly came to mind, but we finished our sauerkraut and went to bed too tired to think about where we were.

After another well-deserved deep sleep, I awoke early, went down to the lobby, walked around the hotel and the big pool and then down to the beach where it was absolutely beautiful, and again, topless! I also found a little hotel just a quarter of a mile away that was our intended and recommended destination, which we moved to later that day. While this little hotel was not right on the beach, with a short walk we were able to use the exact same beach as the German hotel and it also had a nice hotel pool and a quaint restaurant right on the premises. The beach was spectacular, and the healthy German girls were very immodest (to say the least) as they gayly frolicked on the beach!

After enjoying the small, family-run hotel for one day, we were off to Athens by bus. The most memorable scene of that trip was when we crossed the bridge over the Korinthos Canal and watched a huge ocean liner making its way down the canal with what appeared to be mere inches between each side of it and the vertically carved rock faces. The total trip was about four hours, and the scenery was great the entire way with much of the route following the shore of the Mediterranean Sea. Athens was fantastic, such a beautiful ancient city stretching out in every direction with all of its historic architecture highlighted by the acropolis and the downtown Plaka area. The markets were colourful with street musicians and art throughout; the food was delicious and the people very warm and friendly. As this was the beginning of our sixth week of travel, we wanted to find an island where we could relax for several days, so we didn't stay long in the bustling city of Athens; however, I would definitely go back someday to spend more time there. Our next adventure was to an island not far from Piraeus, the main seaport for Athens, where we caught the ferry. It was only a short ferry ride of about one hour to the island of Aegina, where we stayed for several days. The atmosphere on the island was very laid back and casual with lots of nice beaches and outdoor restaurants. This was a perfect place to unwind for a few days as the weather was warm and sunny, the water very warm and the beaches sandy. I remember I rented a scooter on at least one day and toured around the island with either Derek or Taralyn on the back (they took turns). This was a great way to get around the entire island and see all of the sites off the beaten path. One evening, we were all walking around the main harbour area after dinner when a group of young people swarmed us and begged us to come to their hotel where they were having kind of a scavenger hunt and this group's assignment was to find the best mustache on the island! Well, we went along and our group won the contest! I remember the kids thought that was hilarious. Aegina was exactly what we all needed. With its beautiful sandy beaches and warm water, we swam most days. Here was another place where we experienced how salty the Mediterranean Sea was and how much more buoyant we were in it. We could literally float on our backs with our hands up in the air and we were still half out of the water. Well, after a few days of this relaxing and living like millionaires, we were fully recharged and ready for the big trip home.

The day we returned to Athens to fly home was coincidently the same day that President H.W. Bush (#41) was landing in Air Force One in Athens! This was one very big deal with security all over the city like nothing I have ever seen before. Apparently, he was coming to Athens to meet with leaders from surrounding countries to develop a coalition to deal with the civil war then underway in the former Yugoslavia. We planned to go to the airport early that morning, drop off our luggage, then go to a nearby beach and wait for our flight, but it turned out all storage lockers were closed due to concerns about bombs left at the airport. On the way to and from the airport that morning by taxi, we saw literally thousands of policemen standing along the main roads and dozens at every intersection. We did go to the beach with our luggage for a while to kill some time, but we eventually headed back to the airport to get through the massive security and check in. This was a long and slow process due to the heightened security everywhere. Once we got checked in, we went to kill time at a restaurant on the second floor with a full-height glass window that faced the runway, and we luckily selected perfect seats to view the pending proceedings. On and around the airport runway we saw tanks and artillery set up, along with a big bandstand and red carpet running from the taxiway to the bandstand. All kinds of military personnel and police were all over the place, including men with machine guns on all the rooftops in the area. In the distance, we could see warships in the sea just adjacent to the airport. It seemed very tense with this much security going on. Finally, Airforce One landed and taxied over to the area near the bandstand. Then, a second identical Airforce One landed and parked beside the first one. Eventually, the main cabin door opened on one of the planes and George and Barbara Bush came out onto the platform and waved to the crowd right below us. Then, suddenly, a bunch of shooting started! Cannons were firing off from the runway and it was scary, but it turned out to be just a military salute. Shortly thereafter, the Bushes came down the stairs and went onto the stage while the military band played. This went on for a while with speeches, which we could not hear, and then a bunch of black limousines came in and took everyone away, just in time for us to go to our departure lounge. Soon after, we were all stretched out on our flight all the way back to Vancouver, wining and dining in luxury crossing over Greenland, Iceland, the Canadian Arctic, then travelling down the Rocky Mountains and finally into Vancouver, arriving back home on July 21, 1991. This was a once in a lifetime trip for the entire family, and we will have great memories of that trip for the rest of our lives. But now it was time to get back to work!

CHAPTER 19—
The Next 20 Years at WVCC —1991–2011

THE NEXT 20 YEARS WERE quite similar to the previous five years as the company had matured to some degree and found its pace and some of its own niches in the industry. We continued to be competitive in the marketplace, particularly in out-of-town work, which was generally preferred, but some of our field staff may have periodically questioned that philosophy. Average revenues were around $10 million per year, although actual revenues fluctuated significantly from year to year depending on the size and timing of the projects. New bridge construction became less predominant in our workload partially because a lot of highway bridges were now becoming simpler structures in B.C.,[27] but we still did a few of the more difficult and challenging bridge retrofit projects that came out for tender, adding up to just over $33 million over that period. A few significant highway bridge retrofit projects that we undertook very successfully included the Fraser-Hope Bridge, the Churn Creek Bridge near the Gang Ranch, the Harrison River Bridge and the Hyland Creek Bridge near Watson Lake, Yukon. All of those bridges involved retrofitting very old rivetted steel structures and upgrading them with new structural steel, which was quickly becoming a specialty niche of ours. Work for CP Rail

27 This was primarily because of the extended pipe piles we introduced in 1981 in Clearwater. There were now very few difficult foundation jobs for river piers as most bridges had extended pipe piles.

and CN Rail continued up to 1998, with 12 projects worth a total of $11 million, including the extension of an existing snowshed east of Revelstoke, a new rail bridge over the Coquitlam River, the retrofitting of the structural steelwork for the CN Rail Bridge crossing the Fraser River at Prince George, the addition of another rail line along the Port Moody shoreline for West Coast Express and a steel retrofit on the Cisco Bridge near Lytton.

In addition to railway structures, we also completed a number of railway grading jobs managed by Kris Thorleifson, Scott Bradley and Derek Omnichinski, who all joined the company in the mid-90's to take on more earthworks and utilities-type work. Kris started with us soon after our Reed Point Rail Expansion in Port Moody and Scott joined us shortly thereafter, and between them, they started up a separate division for their type of work. Other successful projects completed by this division during this initial period were the installation of sanitary and storm sewers throughout the downtown area of Peachland, which also involved the downtown beautification, and the twinning of Lorimer Road in Whistler.

Hydroelectric projects became one of our new specialties starting with the Soo River Project in 1992, and we eventually built 11 different hydro projects worth a total value of $107,594,700. The largest two were the Miller Creek Hydro Project near Pemberton and the Aberfeldie Redevelopment Project near Cranbrook, valued at $28 million and $38 million, respectively. Hluey Lakes penstock replacement was another hydroelectric project worth $3.3 million up near Dease Lake that was managed by Scott Bradley, with Dale Lauren and Marcel Lefebvre running the work on site. In addition to the above, we also built a few new mines including Kemess South mine in Northern B.C. for over $25 million, QR Gold near Quesnel for $2 million and the Compliance Coal mine near Princeton for $1 million. Although we completed about 85 projects in this 20-year time span, I will only deal with a few of the highlighted projects.

St. John Creek Bridges - Fort St. John, BC

Shortly after I arrived back home from Europe, our crews were just finishing up the Pine Pass Bridges and the final deficiencies at Mt, Hundere Mine, as well as starting to mobilize to our next project, which was the

St. John Creek Bridges. This was a contract with M.O.T.H. with a value of $3,666,000 for four highway bridges near the Beatton River just east of Fort St. John. Art was again selected to run this project as he had just completed his work in the Pine Pass. Each of the four bridges had two large, reinforced concrete abutments up to about 50' high with battered walls and one of the wing walls was composed of a very high battered lock-block wall that required about 750 concrete lock-blocks. The project was designed by ND Lea & Associates and Blair Squire was the project engineer, with Corbett Phillips assisting. The bridge decks comprised of precast concrete I-girders and cast-in-place concrete decks, very similar to two of the bridges in the Pine Pass contract. We set up our Fastway batch plant on site, along with all of our mixer trucks and some local ready-mix trucks we rented by the hour for the deck pours. Carlos Madiera ran the concrete batch plant again to supply all of the concrete we needed. In addition to the ready-mix concrete, we also purchased some steel lock-block forms and produced all of the 750 lock-blocks required for the retaining walls, often with leftover concrete at the end of each pour. One story Art reminded me of during the writing of this book was that they built a wood frame enclosure for the batch plant rather than using a Sprung Structure, and the building nearly burned down one night from the heaters that were used to keep the aggregate warm! I learned at Fort Nelson that wood aggregate bins would also catch fire from the heated pipes below the aggregate piles. (If only the insurance companies really knew what was going on in the field!)

All of the abutment footings were quite deep in order to reach competent bearing soil and avoid scouring by the river so a significant volume of excavation was required as well as the associated trucking to dispose of the excavated material. To perform the excavation and backfill work, we purchased a brand-new Cat 225 DLC, a used Cat D6 dozer and several used dump trucks and self-performed all of the work. I recall that we were trucking the excavated material a couple of kilometers away, and Blair, acting on behalf of the owner and being very cost efficient, started paying us to back-haul structural backfill material supplied by the owner. In some cases, we were building road grade that was outside our contract scope on force account by hauling abutment excavation and then backhauling the structural backfill material. I was quite impressed with this initiative of

Blair's as it made good economic sense and certainly helped us with our costs. Most owner's representatives would not be willing to do that. This was one of our first ventures where we performed all of the earthworks on a project as we often preferred to subcontract this work to competitive local contractors. However, this new policy proved to be a learning curve as the tandem dump trucks were in constant need of repairs and maintenance, so we vowed to never again perform our own trucking with dump trucks. This highlighted one of the general problems in our company: If we could not keep our owned equipment busy on a continual basis with our fluctuating work volumes and differing types of work, it really did not make economic sense to own new equipment; however, in many cases, utilizing used equipment often resulted in excessive repairs, additional maintenance issues and relative cost increases.

Another big aspect of this project was the production and placement of a significant quantity of rip rap around each of the abutments to prevent erosion. As I recall, we retained a local driller/blaster named Vic Gouldie to do the blasting in a nearby rock quarry, then we ran the blasted rock over a steel grizzly (a heavy-duty stationary screen) to produce the required size and gradation for the material. To accomplish this, I wanted to replace our old grizzly and build a brand-new heavy-duty grizzly, and I remember looking at a number of rock grizzlies around the Fraser Valley to get a number of ideas on how best to build one. In the end, we had a fairly stout one built by Benco Fabricating in New Westminster. It could be erected in bolted connections and had big rubber pads built into the framework to cushion some of the shock loading when dumping bucket loads of rock onto the grizzly frame. The screen sizing could also be adjusted to suit the required sizes of rock. It worked quite well, although we had to make some ongoing alterations so the equipment did not self-destruct. Hauling all the rip rap provided plenty of hours for the company-owned dump trucks on site; however, even though they were protected with timber liners, the dump trucks suffered predictable ongoing damage due to the nature of this work.

Art's crew on these bridges included Paul Schlauwitz, Pete Steiner, Wayne Clark, Nick Maskery, Ralph Schlauwitz, Emeric Domokos, Gus Gudmundson and Lloyd McHarg, who ran the 60-ton Koehring crane.

Darryl Unger[28], a former Manning Construction employee also ran the dirt works and rip rap production while that work was underway. Basically, Paul Schlauwitz was the foreman on two of the bridges while Pete Steiner was foreman on the other two. As I recall, we subcontracted the reinforcing steel on this project to RGR Contracting, a collaboration between Bob Ramsay and WVCC. This is where we first met Dennis Case, who was employed by RGR until he came and worked for us directly after this project.

One additional purchase we made for that project was a brand-new, 28-ton Grove all-terrain crane. This was a beautiful machine and so nice to have, especially after the previous 20-ton hydraulic crane we purchased used from Finning turned out to be a total piece of crap. We discovered that all of the electrical contacts were corroded on that crane, causing it to malfunction and require constant troubleshooting and maintenance. We were very near the end of dealing with Finning Tractor at that point!

Once all of the eight cast-in-place abutments were completed, we were ready to erect the precast girders, but it was impossible to get them delivered because they were all locked up on rail cars behind picket lines at Conforce in Richmond. This dispute took almost two months to resolve with literally hundreds of phone calls and mounting legal bills. The girders were eventually shipped from Richmond to North Vancouver by CN Rail in early March 1992, then transferred to BC Rail lines and sent to Fort St. John immediately thereafter. From the rail yard in Fort St. John, we offloaded them from the rail cars and loaded them onto tractors and steering dollies owned by Rocky Mountain Transport. As there was a pretty steep hill on the dirt access road coming down to the site, we had some dicey moments with the girders on steering dollies due to the slippery conditions when it rained, but we made it through without any serious incidents. We erected the girders using Lloyd and the 60-ton crane as well as an 80-ton crane we rented from Fort St. John. Another problem we ran into with the girder erection was when two girders got lost somehow on their rail cars! How does one lose two girders that are 120' long and sitting

28 Darryl was the son of Ike Unger, who we bid the very first job to on the #5 Road Interchange in 1986.

on multiple rail cars? In any event, they were soon found on a siding near Moberly and it took a few additional weeks to get them delivered to the site.

Joe Mather joined the crew for the bridge decks, and the formwork for each of the four bridges was installed in good time as the girder installation was completed. The cast-in-place concrete decks were finished with Clem running the Kingston Construction deck machine again and we were pretty well poured out by June 1, 1992. Darryl Unger and Paul Schlauwitz finished up all of the earthworks and concrete finishing by early July, after Art had left for the next project. Based on the positive things I saw from Blair Squire on this project, on June 2, 1992, I offered him a job to come and work with us as a superintendent, and he joined us a short while later.

Laurie Snowshed - Rogers Pass, BC

We were awarded the Laurie Snowshed project by CP Rail in early May 1992 for the amount of $1,190,000 to extend the existing railway snowshed on the CP mainline along the Illecillewaet River east of Revelstoke. This was a very interesting and challenging project that required working in a very confined canyon location on the active mainline track. This snowshed was situated between a rock face and an avalanche zone above and the raging river below, with very little elbow room between. In order to achieve suitable access to the worksite, we constructed an access road and a two-span temporary bridge over the river, with four used, flat-deck rail cars and a centre pier in the river constructed from rock-filled crib. This was similar to the temporary access bridge I constructed for the McPhee Bridge in Cranbrook many years prior, but in this case, we used old steel beams that we salvaged from the Ashcroft Bridge demolition for the pier cribs instead of timber. This access bridge was designed by Dete Mordhorst at All-Span Engineering to transport loaded highway trucks and concrete mixer trucks, as well as a Manitowoc 4100 230-ton crawler crane that we rented from Lampson Crane for the erection of the precast concrete. Choukalos, Woodburn, McKenzie & Miranda (CWMM) were the project design engineers and Ian Rokeby, who was becoming a snowshed specialist, was our key contact with them. Ian was not only a first-class act and extremely helpful on this project, he continued to be an ally of ours for many years.

Hal Langpap was the site supervisor for CP Rail, while Larry McKee and Al Miltimore were our contacts in the CP head office in Calgary. All of the CP Rail employees were, again, excellent to deal with and even when problems arose, it was always a friendly and collaborative relationship.

Art was assigned to this project as it was near the end of the St. John Creek bridges and he was able to bring with him many of his past crew from his recent projects, including Joe Mather as his general foreman, as well as Wayne Clark, Glen Mosely, Carlos Madiera, Lloyd McHarg, Paul Schlauwitz, Ralph Schlauwitz, Dennis Case and John White. The scope of work for this contract involved demolition of the existing timber shed, construction of new reinforced concrete foundations and the installation of new precast walls and a roof structure. The foundations required detailed rock excavation alongside the existing track for the wall and up on the face of the rock for the footings to support the roof beams. Pressure-grouted Dywidag rock anchors were also installed in each beam footing to resist uplift forces. To provide access for an air track to drill and blast the footings and install all of the Dywidag anchors into the rock face, we constructed a large drilling platform from wide flange beams that Lloyd and the 60-ton crane moved from one location to another. During the first small blast for these beam footings, the crew decided to leave the 60-ton crane in place rather than make the extra effort to move it far away from the blast. Well, you can guess what happened. It got showered lightly with rock, and even though the damage was minimal and superficial, Lloyd wouldn't talk to anyone for a week! That crane was his baby, and thankfully he cared for it like his own. We made some adjustments for the subsequent blasts by using more blast mats and we had no further incidents. At some point later in the job, Glen Mosely and Dennis Case thought it would be funny to put a blow-up doll in the seat of the crane for Lloyd to find when he got to work in the morning. Well, Lloyd didn't find that funny at all, and he went into a rant saying, "You think a dummy can run this crane, is that what you think??" I doubt that Lloyd ever spoke to Dennis or Glen again, at least without a snarl.

Once the reinforced concrete foundations were completed, large precast concrete wall components were erected, braced, and then grouted in place. Next, the sloping roof beams were erected over the tracks, spanning from the footings on the rock face above and cantilevered over the wall. All the components for the precast erection were delivered by railcars to the site and erected by the Manitowoc crane. Once the precast roof panels were all erected and grouted together, we installed a protection board and a complete waterproofing system to the roof of the structure and backfilled the entire shed roof with rocky excavated material that was intended to protect the structure from rock and snowslides. This backfill material was placed with a large mouth concrete bucket hoisted by the 60-ton crane, with a small dozer spreading it on the roof. Again, we batched our own concrete with our Fastway plant that was located close to the access road and run by Carlos and Ralph.

As this was an active mainline, we were given certain hours we could work and certain hours when we were required to cease work and clear the tracks entirely for express trains. We were able to complete the work on time, well before the fall snows arrived and the risk of slides became an issue on the site. Like most of our projects, this was another one with a lot of moving parts and equipment, such as drills, jacklegs, air compressors, generators, loaders and cranes, and it seemed we had more than our share of equipment breakdowns and continuing repairs. We started our company small and had utilized mostly used equipment for a lot of our work up to that point and this practice continued to cause us problems, lots of headaches and probably higher costs as a result.

While we were constructing the snowshed, which was sized accordingly by CP Rail so that we could complete the project in one construction season, we were also negotiating with Larry McKee to perform the next year's snowshed extension under a change order to the contract. This was mainly due to the fact that we could utilize many of the facilities we had already constructed and could leave in place over the winter, including the access road, the bridge across the river and the drilling platforms. Unfortunately, even though CP Rail was keen to have us do the work, they lost their funding for that project, and it was cancelled. I think in the end, we made a deal with them to take over the bridge and other materials for a settlement. We would just have to bid the next shed if it ever came out again.

Soo River Hydroelectric Project - Whistler, BC

The next major project we undertook was the Soo River Hydro project. This was a privately owned hydroelectric development just north of Whistler designed to produce 10 MW of power, which was enough to power all of the Whistler Blackcomb area. As far as we knew, this was only the second Independent Power Project (IPP) in B.C. at that time. The opportunity to provide a proposal to perform the construction of this project came to us through Doug Matheson, a friend of Paul Manning's who had previously worked at Manning Construction. Doug happened to live next door to one of the owner's, Doug Goodbrand (a distant cousin of Jim Goodbrand), and they spoke over the fence about the project. Apparently, Doug was looking for an energetic, resourceful, open-shop contractor that could build the project for his company, Summit Power Corp. Summit was a partnership between Doug Goodbrand, Stu Croft and Tim Sadler-Brown, all very smart and practical guys who had extensive backgrounds in geology, water resources and engineering. They had been awarded a power purchase agreement from BC Hydro to purchase all of the power their facility would eventually produce for 20 years.

Even though we had never built a hydro project before, when we broke it down into individual elements, we didn't see anything we couldn't handle. The project scope involved the construction of a concrete diversion weir located in the upper Soo River to redirect partial river flows, an intake and gate structure to direct the diverted water through a 1 km hard rock tunnel, then the installation of an 800'-long welded steel penstock that was bifurcated through twin turbines in the powerhouse. The de-energized water then meandered slowly out a long, windy tailrace, back to the river below. An associated electrical substation, switchyard and transmission line were also included in the contract. After working closely with Summit Power for many months on construction feasibility, optimization analysis and costing, the contract was awarded in the fall of 1992 in the amount of $9.8 million.

The innovative contract format we negotiated was a guaranteed maximum price, but it was effectively cost plus to a target price with a guaranteed maximum price included to meet the requirements of the lender. Essentially, this was a risk-sharing and collaborative-style contract that encouraged the parties to work closely together to minimize costs while maintaining a specified design and quality criterion. This was the first alternative-style contract format we initiated and we went on to successfully perform many more projects with similar formats in the future. These alternative-style contracts were particularly beneficial on this type of "design-as-you-go" project, for which it was not economically feasible to perform 10–20 years of in-depth analysis and studies in order to finalize the design before commencing construction as large utilities like BC Hydro do on capital projects. (And believe me, they still get it wrong. Just look at Aberfeldie and Site C as examples of that!)

CH2M Hill were retained by Summit Power to be the design consultants and Tom Casher and Tom Field were the two main engineers assigned to the project. Paul Manning worked directly from the site at the beginning of this project, and he and I worked closely with Summit Power and CH2M Hill on all of the design aspects. Blair Squire, who we had hired from the St. John Creek Bridge project, was named as the superintendent, along with Dan Cave, who was assistant superintendent and in charge of quality control. Dan was also assigned numerous studies and analyses, so his nickname on this project became "Study Dan". Again, because of the collaborative nature of the work, we self-performed all of the quality control, and CH2M Hill, who had fairly limited site presence, performed the necessary quality assurance. This was a fairly unique arrangement at that time, but it worked very well and the resulting cooperation between all of the parties was excellent.

We subcontracted the ~10' horseshoe-shaped power tunnel to BAT Construction, who performed the drilling, blasting and rock support as required. Fortunately, 95% of the rock in the tunnel was excellent hard granitic rock, so the tunnel installation went very well. However, there was one weak section of rock with signs of an ancient gravel seam near the upper portal that required some trickier soft tunnelling techniques, but those were also managed quite well. We subcontracted the installation of the steel welded penstock to Graham Wosniuk at Monad Contractors because I had some concerns about our own experience with the handling, welding and coating of the pipe, which I recall was about ½" wall thickness and ~80" diameter. In particular, we had concerns about finding suitable welders who could meet the stringent requirements of the Boiler and Pressure Vessel Code for this work. However, after seeing it done that one time, we self-performed the installation of all penstocks in future projects. The grade of the penstock was 34%, as best as I can recall. All of the electrical work was subcontracted to M. Dickey and Sons from Chilliwack with Bill Soutar as our contact. The transmission line was subcontracted to Pacific Electrical Installations with Mike Bentley as our main contact. The turbine generator installation was subcontracted to Lethbridge Millwright and Welding (LMW), who were very experienced in this type of work. Brothers Alan and Glen Pirot were the owners and main millwrights for LMW.

Again, because we had to place concrete over several seasons as the work progressed, we set up our Fastway batch plant inside a fully winterized Sprung Structure, as I did in Ft. Nelson and as WVCC did for the Pine Pass and St. John Creek bridges. Again, Carlos Madiera assisted with the erection and set up of the plant and he and Paul Schlauwitz batched most of the concrete throughout the duration, using our hodgepodge of older but generally mechanically sound mixer trucks.

Perhaps the riskiest component of the work was the installation of the tunnel through the mountain; however, this work went very well and the rock support costs were kept within a reasonable budget mainly due to the highly consistent and competent rock, as well as the significant tunnelling experience of BAT Construction. Bruce Thompson was the owner of BAT, and he was on site for most of the work. Phil Read was retained by WVCC to supervise the tunnel work and perform rock mapping during its progress. The tunnel excavation was achieved using a typical drill and blast method from each end of the tunnel, and the muck was removed to each portal with typical rubber-tired scoop trams. Ventilation was forced into the head of the tunnel using fans and ducting to provide fresh air and to clear out all smoke, dust and blasting gases. Unfortunately, BAT was suffering through some serious financial difficulties during this work, which complicated completion of the tunnel, but with a lot of effort and handwringing by everyone, it ended up getting done. The massive quantity of tunnel muck (about 20,000 m^3 of blasted rock) that was produced was sold to a developer, and most of this rock was used to convert bog land into the foundation for the Nicklaus North Golf Course and the related housing development, just north of Whistler.

One of the toughest parts of the work was installing the diversion weir across the river, which ran freely, all year long. Partly due to the lack of available subsurface geotechnical information, we were operating somewhat blindly, but we eventually managed to successfully build the weir in two sections by closing off the river in halves with a combination of steel sheet piles, lock-blocks and PVC liner material during the lowest flows in winter. A Bridgestone rubber inflatable dam was then installed onto the completed weir and used to adjust the height of the intake pond during operation of the plant. The operating height of the dam was controlled automatically with a compressor that was run by a Programmable Logic Controller (PLC) based on the available water flows and the capacity of the turbine generators. A number of our regular employees worked on this project including Joe Mather, Pete Steiner, Paul Schlauwitz, Ralph Schlauwitz, Mike Schlauwitz and Mark Stroeder along with Dennis Case and his crew, who placed all the reinforcing steel.

The 800' of 80" diameter steel penstock was installed on the steep, prepared slope in 40' lengths with the use of two Cat 583 side booms (D8 size) running in tandem with a winchline running through a large snatch block that was installed into the rock portal at the top. Welding was performed in place by Monad and coating repairs were taken care of after each field weld was completed. Once the pipe was fully installed, we backfilled the entire pipe with a cement treated backfill material that we called CTB. We sourced fairly clean, natural gravel on site, mixed it with a small quantity of cement in a rolling pug mill, placed it around and over the pipe and then compacted it lightly. After several days, this material hardened up and provided a very solid, durable and non-erodible thermal protection cover for the penstock and still remains to this day, almost 30 years later. The steel bifurcation that split the water into twin turbines was fabricated by Westminster Boiler and Tank and installed by Monad.

The powerhouse was a two-storey building, which was a combination of cast-in-place concrete foundations and precast concrete panels that came with a pre-engineered steel superstructure and roof. The roof was equipped with removable panels to allow future maintenance of the turbine and generator. Joe Mather was the powerhouse general foreman. A pair of 7 MW Francis turbines were installed with matching generators that could be operated individually via a bifurcation in the penstock, depending on available water flows.

In my opinion, the real story about the Soo River Project was not about the innovative design or construction, but more about the collaboration of the people involved that resulted in a very successful project, from both quality and economic perspectives. The beginning of that collaboration was during the negotiation of the contract when we crafted the target price strategy and the risk-sharing principles, not by retaining our own individual lawyers to assist, but in fact, by using one common contract lawyer to our mutual benefit. That lawyer was Ray Schachter, and he was so practical and helpful that both Summit Power and WVCC used Ray on many more projects in the future.

Another great example of collaboration on this job was with Bill Soutar, who was the site superintendent for the electrical contractor, M Dickey and Sons. We met Bill while working on the sewage treatment plant for the District of Chilliwack and really liked working with him, so we invited his company's participation at the Soo River Project. In my opinion, he was a genius when it came to finding the most practical and cost-effective ways of finalizing the electrical design, and he seemed to be far more motivated by coming up with better and simpler means of completing the work than making more money as a contractor. Because we were participating in a cost- reimbursable-to-a-target-price style contract, we and the owner loved working with Bill and appreciated his design and construct strategies. In fact, we liked working with Bill so much that after the Soo River Project was completed, we approached him to start our own electrical company together, and that was the beginning of Westpark Electric.

Like any industrial, greenside construction project, there were surprises and problems that arose, but we collectively came to practical solutions on each of them and the entire project came in under the target budget with both parties sharing equally in the cost underrun. Another hidden bonus was that although the project was originally designed conservatively to produce 10 MW, it has been successfully operating at 14 MW for many years. Soo River, while admittedly being "low hanging fruit" at the beginning of the development of these IPPs in B.C., might be the only project that exceeded the design capacity and actually underran in costs. Stu Croft, Doug Goodbrand and Tim Sadler Brown have remained good friends with us ever since that project, and we have often fished together at Hakai Pass[29] on the central coast, as all of us are fishing enthusiasts.

Soo River Hydro - Tailrace, Powerhouse and Buried Penstock

29 Hakai Land & Sea is a private, land-based, fly-in salmon fishing club on Calvert Island up on the central coast of B.C. This club was founded in the 1970's by a number of businessmen including Dave Manning, and it continues to operate to this day. Paul Manning was one of the directors of the camp for many years and Paul's cousin Ben Angus runs the camp every summer. I was fortunate enough to join this club around 1988 and have been going there every year since then with many guests and clients of WVCC. The facilities of the camp have been dramatically improving over the years and the experience is more enjoyable every year I go!

Bonnington Falls Generating Station - Nelson, BC

Another interesting hydroelectric project we started in 1994 was the Unit 5 replacement at Bonnington Falls for the City of Nelson, which was a contract worth $3.7 million. The city had been operating several powerhouses on the Kootenay River approximately midway between Nelson and Castlegar, with their first unit going into operation in 1905. This new facility replaced a very old, yet still operating, powerhouse that, remarkably, used wooden wedges as the main bearings for the vertical turbine shaft. That powerhouse continued to operate right up until the time we completed our work. Peter Hartridge was the city's representative who I dealt with and Chuck Stuart was the resident engineer on the site. EBA Engineering was the prime consultant for the project and Ian Stewart was the lead engineer.

Our project specifically involved the construction of a concrete intake gate structure at the head of a tunnel and a complete new stand-alone sub-terranean powerhouse. The short tunnel was constructed by a local contractor just prior to our work getting underway at each end. Both the intake structure and the powerhouse included some very difficult transition formwork between the square gate at the intake and the round shaped tunnel, as well as between the turbine scroll case and the square outlet gate in the powerhouse. Also, the new powerhouse was carved vertically out of rock and most of the high wall formwork was one-sided wall forms that were anchored into the rock faces. Art Penner was the superintendent of this project and Joe Mather and Pete Steiner were the carpenter foremen. Dennis Case, who had been with us for a couple of years, placed all of the reinforcing steel and assisted with the installation of the large embedded items that the turbine generator supplier provided to us. Again, we set up the Fastway concrete plant on site and produced all of the concrete for this project with Paul Schlauwitz running this operation. Tebo Mill Installations was subcontracted to supply and install the various mechanical components in the powerhouse, and they also installed the owner-supplied turbine generator equipment, under the direction of Voith

Hydro, the owner's equipment supplier. Martech Electric, out of Castlegar, performed all of the electrical work under a subcontract to WVCC as well.

Bonnington Falls Hydro - Looking Down Into the Powerhouse

Running concurrently with Bonnington Falls, we were awarded another project in the amount of $1.25 million from West Kootenay Power on the adjacent dam just upstream on the Kootenay River, called the Corra Linn Dam. The scope of work of that project was to jackhammer the downstream face of the entire concrete dam down to the reinforcing steel and then replace all of that concrete. Bill Campbell, who was a former co-worker of mine at Goodbrand and the Conveyor Crossing job, was the superintendent of that project, and we delivered all of the ready-mix concrete from our batch plant at Bonnington Falls.

Corra Linn Dam

While I consider the Bonnington Falls project to be a fairly successful one, several unfortunate incidents occurred during and after construction that are story-worthy enough and should not go unmentioned. The first incident involved a lot of problems we encountered with the steel intake head gate that was fabricated and coated by the H. Fontaine company near Montreal, Quebec, and then shipped to us for installation. These steel fabrication and coating problems caused us delays and additional costs due to what seemed to me to be lack of experience by H. Fontaine in these types of high-pressure gates. It seemed their gate manufacturing expertise was perhaps more suited for lower-head, agricultural uses. As problems related to the fabrication and coating defects arose, the more wary and particular the design engineers became, leading them to continue to discover more issues, resulting in some serious cost and delay impacts. These delays in completion resulted not only in our own increased indirect costs but also liquidated damages of $5000 per day assessed against WVCC from the City of Nelson, complicating matters even more. H. Fontaine was not fully cooperative in rectifying these defective issues and a dispute between us arose to a point where both Paul and I flew to Montreal on November 10, 1994, to meet with the Andre Fontaine, who was the owner and manager of the company. Our purpose for the meeting was to convince H. Fontaine that our issues were serious and that we would end up in an arbitration process if they failed to repair the gates to meet the contract specifications and to take responsibility for the delays and the resulting liquidated damages. Unfortunately, our advice fell on deaf ears. As disappointing as the results were from this meeting, Paul and I both loved Montreal, and we made the very best of our short time there, enjoying nice restaurants and deluxe accommodations in Old Montreal.[30] We learned that day that November 11 is not celebrated in Quebec as it is in B.C., so the bars and nightclubs were packed until 3:00 AM! (Or so Paul told me anyways!)

30 Somehow Paul ended up getting two hockey tickets to a Montreal Canadiens game that night in the old Forum so that was pretty cool to experience that before the Forum was replaced. Another great experience was being in the mob scene as everyone walked out of the Forum after the game and straight down the busy downtown streets where it seemed everyone was heading to a variety of packed bars and nightclubs.

Since the City of Nelson was assessing liquidated damages to WVCC for the delays caused by H. Fontaine, there were effectively three parties involved in the dispute, causing us to commence a tri-party arbitration process in order to resolve the dispute in accordance with our prime contract with the City and the subcontract with H. Fontaine. At this point, the issue involved many tens of thousands of dollars and Bob Jenkins (who was a founding partner of Jenkins, Marzban, Logan, a well-known construction litigation firm in Vancouver) was selected by all three parties to be the arbitrator, hear the case and provide his binding judgement. Mike Demers was a young construction lawyer with Owen Bird, and he was assigned by that firm to represent WVCC in this dispute, as all of our legal affairs were handled by Owen Bird at that time. In short, Mike performed extremely well, but due to the confidentiality of the arbitration process, I unfortunately cannot elaborate further.

The next unfortunate instance involved a flood within the powerhouse during the installation of mechanical equipment by our mechanical subcontractor, Tebo Mill Installations. This flood required an extensive cleanup and damaged some of the mechanical equipment that Tebo had installed, including the HPU (hydraulic power unit), which had to be completely dismantled and sent out for a full flushing and reconditioning. The various repairs ended up causing significant additional costs and critical path delays to the project, and these delays then triggered additional liquidated damages assessed to WVCC from the City, and again, created another tri-party dispute. Tebo's position was that WVCC was somehow responsible for the flood, and they refused to take responsibility for any of the resulting delays or costs, even though it was clear to us that the flood resulted from the failure of their sump dewatering pumps that were already installed, commissioned and in full use. Again, this issue involved many tens of thousands of dollars and Tebo remained firm in their position, so WVCC had no alternative than to take the matter to binding arbitration for final resolution. Because the City was again involved, the parties again agreed to utilize the same tri-party process and arbitrator as in the H. Fontaine matter, under the terms of the prime contract and the subcontract with Tebo.

Mike Demers again handled our case extremely well and it was apparent that the arbitrator, Bob Jenkins, was also very impressed with Mike's knowledge and tactics used in the hearing. Soon after, Mike moved his employment from Owen Bird to Jenkins Marzban Logan, where he would work for many years. Again, due to the confidential nature of the arbitration process, I cannot elaborate on the outcome of this case, but WVCC immediately became a client of Jenkins Marzban Logan, and we worked with Mike many times over the next 20 years.

The final unfortunate incident on this project that requires mention was a major explosion and fire in the powerhouse. This incident occurred just slightly over 12 months from when we achieved final completion under the terms of the contract and the warranty specified in our contract was limited to 12 months. The explosion was a very serious one that put the entire plant out of operation for quite some time, yet we were hopeful, due to the warranty provisions, that neither WVCC nor its subcontractors would be involved in what would likely be a long and drawn-out investigation, reconstruction and (most likely) litigation. However, nothing is that easy and our insurance company, its agents and lawyers were involved for quite a while. As it turned out, we were not found liable as it was (at least anecdotally) determined the cause of the explosion was due to a design error in the overall plant's electrical logic, something that was the responsibility of the City of Nelson, their electrical design consultant and their respective insurers. It was later determined (anecdotally) in the investigation that when the plant operator was trying to shut the plant down for maintenance, the main generator that is run by the turbine to produce electricity, somehow turned into a motor and began drawing electrical energy from the grid causing it to overspeed and eventually explode. This definitely should not be able to happen and the safety aspect to avoid such a predicament was somehow missed (anecdotally) in the electrical design. Other than these unfortunate issues, the Bonnington Falls Project was a very interesting and enjoyable project, and it was also fun to spend time in Nelson again, which is a great spot to live and work.

QR Gold Mine - Quesnel, BC

The QR Gold project was another brand-new mine that we built near Quesnel, B.C. for Kinross Gold Corp. in 1994–95. The value of that project was just over $2 million and involved all of the civil concrete work for the entire mine facility. Merit Consultants was again the consultant and Eric Smith was the on-site construction manager for Merit. Blair Squire ran that project for WVCC and had Ed Brandl as his surveyor/assistant superintendent. Blair hired some crew he had worked with before who were from Newfoundland (Okay, they were Newfies.), and they collectively broke records with their very high man-hour productivity. Ronnie McGinnis and Peter Morrissey were his carpenter foremen, Tim Fitzgerald was his labour foreman and the rest of the crew were mostly local tradespeople. Blair selected both Ronnie and Peter as his carpenter foremen, and I apparently took some issue with that because both of them would be paid the top hourly rate at that position. As we were open-shop, we wanted to pay workers based on merit, but we also had to be careful not to have everyone working at the top wage. Nevertheless, to Blair's credit, he convinced me, and those guys turned out to be perhaps the best tandem of carpenter foremen I ever saw, well worth every penny of their wage!

The scope of our work on this project involved the construction of the concentrator building, including the ball and sag mill foundations, the crusher, the coarse ore facility and the service complex. We again erected the Fastway batch plant and produced about 2600 m^3 of concrete for the project, with a unit cost of $135.00/$m^3$ including winter concrete. As this project was again built over the winter, the plant had to be fully winterized, and all areas being formed required extensive hoarding and winter heating. One story that Blair reminded me of during the writing of this book was how he dealt with the ground frost in the mill building. The work started in the fall with the construction of all perimeter footings and walls along with various building and equipment foundations throughout the building. By the time that work was complete, there was at least 5' of frost in the ground throughout the entire building and we had to build the slabs on grade. As it was Christmas break at that point, Blair had the crew build a low tent over the entire mill building, then placed a 1 million BTU propane heater in each corner under the tent and let them run over the holidays, with a crew checking them day to day. When the crew returned in early January, the frost was completely gone and the crew got right to work on the unfrozen ground!

I think this was about the seventh mine contract we'd performed for Merit Consultants and because the project went so well, I am sure Merit greatly appreciated the great work that Blair and his crew achieved. However, like the Wathl Creek Bridge, it was one of those projects where the man-hour productivities were so good you had to be careful of using them for future estimating.

Fraser Hope Bridge Reconstruction - Hope, BC

This was another award from M.O.T.H. in 1995 in the amount of $5.7 million for the complete refurbishment of the superstructure of the existing highway bridge over the Fraser River on Highway 1 in Hope, B.C. Art was again named the superintendent on this project, and he had many of our regular guys on his crew, including Clem Buettner, Dennis Case, Pete Steiner and Paul Schlauwitz. Wilfred Fry was the district manager for the Ministry and Boris Obrknezev was the Ministry's project supervisor on site. As I recall, Terry McKay supervised the steel retrofit work on site for the Ministry. The bridge was originally built in 1915 by the Kettle Valley Railway (KVR) as a double-deck bridge with the railway going through the steel trusses and automobiles travelling on the top deck. The bridge was approximately 1000' long and fabricated using rivets and built-up steel sections, as they did in those days before bolted connections and larger rolled, wide flange sections became common. The railway ceased to operate over this bridge in 1959, but the top deck continued for vehicular traffic, although it was narrow and constructed of open steel grating. This contract required extensive reinforcement of the existing structural steel supporting the upper deck as well as installation of one additional stringer on each edge of the superstructure to widen the deck. Due to the requirement to keep one lane open throughout our construction, we performed most of the work in one lane at a time and worked two shifts for much of the work to shorten the overall schedule. Clem generally ran the night shift.

KWH Constructors was again awarded a subcontract to remove, replace and install all structural steel members. Bob Hawk was the project manager and Brent Johnstone was the site superintendent for KWH, and he performed very well. The steel refurbishment involved removing thousands of rivets using pneumatic rivet busters and then replacing those rivets with high-strength bolts. Most of that steelwork was required beneath the top deck, so access to that work was easily provided by the lower deck where the railway once operated. KWH was again operating under a target price contract where we cooperated fully to complete the work as quickly and efficiently as possible. One other aspect of this bridge project was that we were required to "rent" lane closures whenever we needed to close a traffic lane to complete certain portions of the work. This was intended to reduce the overall impact to the travelling public. Reducing the amount of lane closures led to lane rental savings, which significantly reduced our estimated costs. These savings were again shared between WVCC and KWH due to our cooperative contractual arrangement. Once the steelwork was completed in a lane, we followed behind, constructing a much wider reinforced concrete bridge deck along with a sidewalk, new concrete parapets and railing. One mishap occurred on one night shift when KWH was erecting the very last stringer in one of the lanes with a local rented hydraulic crane. Even though KWH had the most stringent safety and engineering procedures in place to avoid this type of incident, someone failed to follow the exacting procedures simply because the final stringer was only 30' long compared to the rest of the spans, which were 60' long. The crane tipped over, the boom landed on the parapet of the traffic lane and the stringer fell into the river, never to be seen again. Luckily, no one was hurt and the crane suffered only minimal damage, but the traffic lane had to be shut down for the rest of the night until the crane could be righted in the morning. KWH scrambled to get a replacement stringer fabricated and eventually finished the steel erection soon after this incident.

On February 7, 1996, while we were partway through the project, I received a call from Wilfred Fry. He wanted us to consider a proposal to investigate and provide pricing for a fairly major change to the design of the bridge deck. This change included elimination of the waterproofing membrane and the asphalt overlay and the installation of a thicker concrete deck using "high performance concrete". I think this change may have had something to do with some previous failures of the waterproofing membrane on bridge decks, including a very serious delamination failure on the Oak Street Bridge in Vancouver. High performance concrete was a fairly new term used in those days, but basically, it was high-strength concrete with very low permeability. These qualities were achieved with a very low water to cement ratio along with the use of silica fume and the addition of high doses of superplasticizer to greatly improve workability. We priced the change order to the Ministry and they accepted our terms. We worked closely with Phil Seabrook at Levelton & Associates who was a great help in getting this concrete mix design finalized and a test pour completed, a request by the Ministry to see how the new concrete finished with a deck machine. The test pour was an actual real-life mock up during which we placed a section of deck in the yard at Hope Ready Mix with the deck machine, and although everything went very well, we learned a few things and made some adjustments. We eventually placed the entire bridge deck with this new mix design, and it went extremely well. Another difference with high-performance concrete is that it is placed at a reasonably high slump and then rough-tined, as opposed to being placed with a very low slump and applying a light broom finish, which makes the finishing quite easy. The one concern with this mix design was the potentially higher heat of hydration and the potential associated thermal cracking, so it was very important to wet cure the deck as soon as possible and to maintain that curing for a full seven days. This bridge deck is now 25 years old, and there is virtually no sign of any cracking and the deck surface is smooth and wearing extremely well.

The Fraser Hope Bridge was another good example of a job going very well. We enjoyed outstanding supervision from Art as well as excellent cooperation from all of our subcontractors and the Ministry staff. I can assure you this was not the case on many projects we undertook through the mid-1990's. My review of the notes of that era revealed I was constantly swamped with subcontractor disputes like the two at Bonnington Falls,

Victoria Machinery Depot suing us over the steelwork supplied for the Ashcroft bridge, several arguments with the District of Chilliwack over some sewage treatment plant and pump station work we completed and a fairly minor crane accident that occurred on a small CP Rail job that ended up in court. One other dispute arose on another project for CP Rail when a hired electrical subcontractor drove a ground rod for an office trailer through a fibre optic line and BC Tel sued everyone in the telephone book, including us. All these disputes were eventually resolved quite favourably for us, but every one of them took a lot of time and effort and drained a lot of energy from myself and the company. Therefore, we made a concerted effort over the next number of years to avoid these confrontations, almost at any cost. As pleasurable and rewarding as it was to win most of these cases, I decided it would be far better and more productive to avoid these situations as best we could. They definitely added a lot to my workload and stress.

From 1993 to 1995 inclusive, we were awarded 11 projects with a total value of $16.3 million, averaging relatively low at only $5.4 million worth of new projects per year. A major reason was our involvement in the Soo River Project, which was a very involved, multi-year undertaking. Contrasting this period to 1996, we were awarded a total of nine projects with a total value of $31.6 million. This included a nice, big, new mine worth $25.3 million to us: the Kemess South Mine.

However, before we jump to that project, one job of several that Kris Thorleifson and Scott Bradley managed for us around that time was the North Bend Siding Extension, which added additional rail lines to CP Rail's existing siding across the river from Boston Bar. This was a straightforward earth cut with 500,000 m^3 of dirt to excavate and haul to construct a section of new grade. The remainder of the dirt was hauled to a dumpsite about 1 km away. This was a beautifully orchestrated job of high production and the cheapest dirt moving I have ever seen. Rusty Dodds was the on-site foreman who Kris hired from previous work with him and Hal Langpap was again the representative on site for CP Rail. Rusty was a crusty old sort who had a large red nose and was not likely the most politically correct foreman of the day, but boy, he knew how to move dirt! Our fleet of equipment included a Hitachi 400 excavator, a Komatsu 300 excavator, a D8 Cat, seven 30-ton Volvo trucks, a D6 Cat and a ride-on packer. All of this equipment, except some of the Volvos and the packer, was company-owned equipment as our overall fleet had been growing steadily since about 1990. The cut was about 50 m high and probably ½ km long, so the 300 excavator performed the trimming at the top of the cut with the D8 Cat pushing material down the slope to the 400 Hitachi excavator loading at the bottom. Gord Trottier operated the big hoe at the bottom, loading the trucks with our biggest cleanup bucket, and he filled each truck in about 30 seconds with four fully heaped buckets. By the time the next truck backed in to be loaded, he had repositioned his excavator and had another overflowing bucket in the air and another truck was gone in no time. There was never a wasted move or seconds lost with Gord. Gord, who came from the Goodbrand days in Hope, was the finest production hoe operator I have seen. He continued to work for us on several other jobs, usually on very steep and dangerous work or where we needed very high production. I visited this site at least once and it was amazing and very pleasurable to watch this kind of production. Rusty's job was to keep all of the equipment working efficiently, keep the trucks moving and ensure there was no let up. (We handled all of the employment standards and labour relations issues after the dirt was moved!) Over the life of that project, we moved an average of 500 m^3 of dirt per hour or 5000 m^3 per 10-hour shift with a unit cost of $1.00/$m^3$ for excavation, including slope grading and loading the Volvo trucks. The entire crew had to produce at a very high level of efficiency to achieve that production.

The interesting thing was that our contract with CP Rail was on a cost-plus basis, so we only made a markup on the total costs, but I can assure you that CP Rail loved Kris and Rusty for those unbelievably low costs! When that job was completed, I bet the bars in Boston Bar were hit with an economic recession because as hard as those boys worked during the day, I am sure they made up for it at night!

Kemess South Mine - North-Central BC

The Kemess South Mine was a large new mine development owned by Royal Oak Mines with a budget of $390 million. The project was located in

North Central B.C., approximately 450 km north of Prince George as the crow flies, near the head of the Findlay River. The Omineca Resource Road was a gravel road about 550 km long from Prince George to the site with highly variable conditions, particularly in the beginning when this work was just getting underway. The road was constructed for logging trucks, and although I never took the opportunity or the time to drive that route, I understand there were lots of rough spots, making it a long, arduous route for our many transport trucks. The mine owner constructed an airstrip on site serviced by air out of Smithers, with a flying distance of 280 km. The copper-gold mine was designed for 50,000 tonnes per day through-put, which was far bigger than any other mine we had been involved in, so the quantities of our work were also massive. The contract value of this primarily civil work package was $23.5 million. The scope of work for this contract included construction of the foundations for the concentrator building, a large service complex building, the crusher building, coarse ore tunnels, numerous flotation tank bases and all the other ancillary structures on the mine site. We had a second contract with Royal Oak in the amount of $1.9 million for the supply and installation of all associated yard services, including all potable and mill wastewater services, a complete sanitary sewage system and a small sewage treatment plant. The crusher building included a huge 104-foot high Hilfiker wire mesh retaining wall with a total face area of 24,000 SF. That was the highest near-vertical, mechanically stabilized earth retaining wall in the entire world at that time, and it was designed to support the mine's fully loaded Cat 793 haul trucks at that height.

Kemess South Mine Crusher

Blair Squire was the project manager for this job. He worked with many long-term key employees including Clem Buettner and Bill Campbell as superintendents; Joe Mather, Pete Steiner, Paul Schlauwitz, Denis Casavant as foremen; and Paul Bradford and Dale Lauren as two newly hired assistant superintendents. Ron Brown was one of the surveyors on this job. Derrick Omnichinski was the superintendent responsible for the yard services contract. We mobilized to the site in late summer 1996, and we commenced work early that fall. Jim Koski was the construction manager for Royal Oak and John Rempel, Doug Anthony, John Barnett, Peter King and Marc Hewison were his assistants. Jay Collins was the main representative from Merit Consultants, who again acted as construction manager. We also worked closely with Johnathan Clegg, Alain Catteau and Terry Shepard from Kilborn Engineering. Golden Hill Ventures (GHV) was awarded a contract from Royal Oak to perform the site clearing, mass excavations and the construction of the airport, so most of their work preceded ours. However, right from the beginning, Royal Oak was way behind schedule and very unprepared and understaffed to get such a large project underway. From a review of my notes during that start-up, it was apparent that Royal Oak didn't even have a scheduler hired on site by the end of November, so it became very difficult and near impossible for the various contractors to plan their work, work that was very often highly interdependent with other tasks. In fact, Royal Oak did not get cutting permits issued from BC Forestry for parts of the work until January 1997, seriously delaying the start of a lot of GHV's work. Various site work was actually shut down in January due to a number of serious delay issues on site. The beginning of this project was very difficult and the access road to the site was becoming one of the major factors. The monumental task of getting all our materials and equipment shipped to site became a logistical nightmare. All in all, we had approximately 30–40 loads of construction equipment, 125 truck loads of rebar, over 500 truck loads of bulk cement and bagged fly ash as well as numerous loads of miscellaneous metals, dozens of loads of Hilfiker wire mesh and corrugated steel pipe, not to mention many loads of lumber, tool containers, lunchrooms, offices, etc. In addition to these early mobilization issues, we had to endure very difficult camp conditions at the beginning of the job, including room shortages that caused several men to live in each room, food shortages and poor-quality meals, seating shortages in camp preventing everyone from eating at one time, as well as having no telephone or television. These issues were a recipe for disaster at the beginning of such a large, complex project of this nature in such an isolated location. In the meantime, Peggy Witte, the owner of Royal Oak Mines had a complete modular home installed onto the property so she would have comparatively luxurious accommodations if and when she made an appearance at the site.

Clem Buettner was the crusher superintendent along with "Crusher Joe" Mather as the crusher foreman. Bill Campbell was the superintendent of the coarse ore facility and Paul Bradford was the superintendent of the service complex. Pete Steiner and Denis Casavant took on the huge job of building the concentrator building, including all of the perimeter foundations, walls, tank bases and the massive grinding mill foundations and pedestals. The grinding mill foundations alone required over 4600 m^3 of concrete and the grinding mill pedestals were 30' in height with over 3050 m^3 of additional concrete. Dale Lauren, a recent BCIT engineering grad, was very experienced for his age and was assigned to assist Blair in all areas of the project. Dale proved to be a valuable asset to us on that job and for many years to come.

The total volume of concrete required for our contract work was 28,340 m^3, and we site batched all of this concrete with our brand-new Ross ready-mix plant that we purchased out of Texas. Art reminded me during the writing of this book that he drove up to the site in a convoy of our mixer trucks with Lloyd McHarg, Dennis Case and Paul Schlauwitz, towing our mobile Fastway batcher to build the foundations for the Ross plant before any of our other work was underway. Art was working on the design and layout for the new concrete plant and his crew poured all of the concrete foundations for the plant, drilled a water well, set up the Sprung Structure that we owned for heating the aggregate and set up a new, larger Coverall building that we purchased to house the plant and all of the mixer trucks. We also had a split aggregate hopper and a conveyor belt that ran from the Sprung Structure to the charging bins of the plant in the Coverall. As this project would operate year-round, we set up the chiller to cool water in the summer and a boiler to heat water in the winter. Two things that both Art and Paul remember clearly were the terrible, substandard camp conditions (the crew lived in four-man wall tents before the construction camp was even set up) and the poor condition of the access road. Paul Schlauwitz ran the plant for us again, and we had all six of our fully refurbished and newly painted mixer trucks and our Cat 966 D mobilized to this site. We purchased a Schwing 32-m concrete pump truck from Joe Delehay at Surrey Concrete Pumping to place all of the concrete and Joe loaned us a full-time pump operator to run it for the duration of the project. I think we purchased the pump for $350,000 and sold it back to Joe at the end of the project for $300,000, so it basically cost us $1.76/m^3 for the pump, plus fuels, maintenance, etc., which was pretty cheap! Overall, the volume of concrete overran the original estimated quantity by 25% due to the many changes in final design. This was great news because our own operating costs for the batch plant operation were under budget in the range of 35%, which added up to a huge savings. (This was a savings of about $200k in labour alone. We owned all of the equipment involved in this operation, so again, this concrete production was very lucrative.) Once the project was completed, we sold off the Ross concrete plant to Inland Cement in Alberta, so the net cost per cubic meter of concrete for this brand-new equipment was a mere few dollars.

Dennis Case was again our rebar foreman, and he and his crew placed a total of 2600 tonnes (5.7 million lb. for you old guys) of reinforcing steel on that job for a net average of nine man-hours per tonne, while we had a budget of 10 MH/TE, which was a net savings of over $130k. Dennis, who was an ironworker foreman we met on the St. John Creek Bridges, came to WVCC in 1992 for the Soo River project and turned into a goldmine for us as he was a very smart, multi-talented and productive worker. Because of Dennis, we were able to perform all of our own reinforcing steel installation, which was a very valuable asset on those out-of-town projects. His versatility was also extremely valuable—he was an excellent rigger and he became an outstanding mixer truck driver. Dennis was the epitome of a great open-shop worker because of his multiple talents, his drive for production and his popularity with all of the workers. In later jobs, we also took on most of the structural steel retrofits and steel erections because of Dennis's abilities and experience in those fields.

Lloyd McHarg was again on site running the 60-ton Koehring crane and Gus Gudmundson apprenticed under him on that crane and on our 28-ton Grove hydraulic crane. The rest of the crew included many of the regulars: Mike Dorris, John Jaconen, Russ White, Blaine Benney, Wayne Clark, Glen Mosely, Ralph Schlauwitz, Korey Mather, Gerry McMillan and many more. Derek Omnichinski was superintendent of the yard services contract, and he had three excavators on site to install all of the site piping as well as the permanent sewage treatment plant. Derek reported to Scott Bradley and Kris Thorleifson, who continued to manage our earthworks and pipe division from our Langley office.

Late that fall, we had two terrible incidents occur on site. On November 27, 1996, Clem had another unexpected heart attack while he was working at the site. He was treated immediately by the camp nurse and stabilized with a plan to medivac him out of the camp to the hospital in Prince George; however, it was nighttime and snowing, so helicopters were not able to get in under those conditions. Apparently, the runway was also not equipped for fixed-wing night flights of that nature. His condition rose and fell throughout the evening, but he eventually passed away with the camp nurse doing everything possible to help him. I still recall Blair telling me that Clem maintained his humour right up to the end, kibitzing with the beautiful camp nurse. Clem's untimely passing was absolutely devastating to Blair and the entire crew on site, as Clem was a very popular man. I was personally devastated as not only was Clem a great and dedicated employee, over the last 19 years he had become one of my best friends. Blair kept me advised by satellite telephone through the evening right up to when he passed, and when I heard the phone ring in the middle of the night, I knew it was not a good thing. Clem's funeral was held in Clearwater and literally dozens of co-workers attended from all over B.C., along with his family and many friends. It was a very sad occasion as Clem was only 57 years old. There is a brass plaque erected in the crusher building of the Kemess mine site in memorial to Clem.

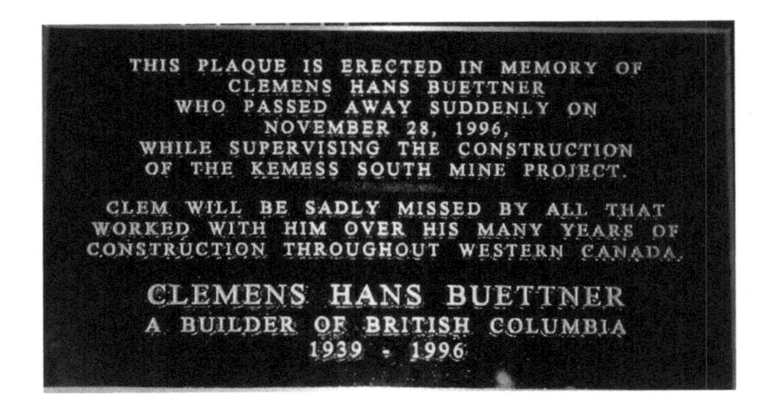

Clem's Brass Plaque

Clem's Crusher Crew

The other unfortunate incident was an accident with our 60-ton crane that occurred on site December 9, 1996. The operator trainee (It wasn't Lloyd!) accidentally pulled the headache ball up through the boom tip at full throttle, which collapsed the entire boom, causing it to fall down mangled onto the crane itself and into the work area. I don't believe there were any injuries, so that was very fortunate. The damage to the crane was about $75,000, and the repairs required the complete crane and boom to be shipped down to Vancouver, which was no small task considering the remote location. These significant repairs were not completed until the end of March and that tied me up significantly with all kinds of engineering, materials testing and paperwork to get the boom rebuilt and recertified.

One close call that Blair reminded me of during our discussions while I was writing this book occurred on this job during the concrete pour for one of the four massive ball mill foundations, which had volumes of about 1,000 m^3 each. Several hours into this particular pour, Pete Steiner got hold of Blair by radio and told him that some of the high-strength coil connectors were snapping, many of the form plates were well sunken into the walers and it looked like the formwork was going to explode. Once Blair got there, they determined they had been pouring too quickly and the form pressure was way too high! They agreed to shut the pour down for a couple of hours and let the concrete start setting up before recommencing the pour. This situation can sometimes occur when fly ash is used in the mix design in cold winter temperatures and the concrete doesn't set up as quickly as normal even though the aggregate is heated and hot water is used for batching. In the meantime, during the delay, they roughed up the construction joint so that a good bond could be achieved when the pour was restarted. After a two-hour delay, they started placing concrete again and eventually completed the foundation without further issues. Once the forms were stripped, there was no evidence of the cold joint or any deflections in the finished concrete. That was a very close call. During one other ball mill foundation pour, the ambient temperature was -25°C and the pump line froze up in the middle of the pour due to a short delay in delivery. Luckily, they had the 60-ton crane set up with concrete buckets to carry on the pour until the pump line could be thawed with tiger torches and propane heaters. Meanwhile, Pete set up a temporary hoarding over this pour and poured on the heat big time to keep everything from freezing up. (There are 1000 stories about working in the cold!)

Working floors, or mud slabs, as they are referred in the industry, are thin, unreinforced concrete slabs, maybe 50–75 mm thick used to level ground and provide neat, clean areas on which to perform accurate layout, place rebar and commence forming footings or foundations. Our contract stipulated that mud slabs were to be paid by m^2, and we had a good unit rate to install them. Doug Anthony, Blair's contact with Royal Oak, often argued with Blair that they were an unnecessary expense, but Blair apparently convinced him that mud slabs improved the quality of the work and critical path scheduling. Hence, we placed mud slabs all over the entire concentrator building, often very cheaply by using waste concrete left over from our concrete operations. This was another lucrative component of the job, but to Blair's credit, mud slabs are also very useful and helpful for quality, workmanship and production.

One other issue that arose in the spring during breakup was that as the frost came out of the ground, the long access road started to self-destruct and significant road restrictions were imposed by the Ministry of Forests on loads to site for a couple of months. This is a regular occurrence during breakup in the north and should be expected; however, it almost seemed that the owner was caught unaware of this serious situation and was unprepared to provide adequate maintenance to the road; yet, at the same time, they could not afford any delays to construction. Due to the conditions of the road, we were eventually paid to haul cement and fly ash in large sacks on partly loaded highboy trucks because the bulk tanker trucks could not haul any product without exceeding the limits of the road restrictions. For a period, when sections of the access road were effectively inoperable, the owner arranged to put up to 20 loaded trucks on several flat barges and tow them up to the end of Williston Lake. From that point, the trucks were able to drive to the Kemess mine site with partial loads and then the empty trucks were able to return south on the access road. Getting trucks onto the barge became a bit of a lottery as every contractor on site at that time was in the same boat as us and badly needed materials and equipment to be delivered. Unfortunately, this occurred right when our loads of Hilfiker materials were being shipped from California, so this definitely added some complexity to the process of us receiving that material in the correct order and as we needed it for installation. While these efforts kept the project going to some degree, this entire barging exercise caused a lot of logistical issues and a lot of extra costs, which the owner ultimately paid for.

Overall, the Kemess Mine project was a very successful one for WVCC. In some cases, we made money the old-fashioned way by achieving higher productivities than estimated and making good deals in our business decisions, but in many other cases, we made great profits by overruns in the right quantities of work. Examples of this were increases in rebar quantities and concrete production, where we were already significantly underrunning the budgets in those items. However, other changes resulting from final design drawings compared to the original tender drawings were also very profitable. One such example was where the formwork to concrete ratio changed in some areas, which was important because all formwork was paid incidental to the concrete volume. One simple example of this change was the grinding mill pedestals, where the formwork to concrete ratio fell from 1.92 m^2/m^3 to 1.29 m^2/m^3, which resulted in a reduction of budgeted formwork of 1907 m^2. Based on the labour and material budget for the formwork built into the unit price for concrete, that was a single savings of over $200,000. Again, the fact that the mill pedestals' concrete volume almost doubled from 2019 m^3 to 3949 m^3 certainly didn't hurt either.

As our work started to wind down, we became more and more concerned about receiving prompt payment from Royal Oak as they were falling seriously behind in their monthly progress payments to us. We soon learned we were not alone in those concerns as it turned out all of the contractors on site were having similar issues. As I became aware of the severity of this problem, and due to the fact that we were owed approximately $8 million in overdue payments by that time, I decided to call each and every other contractor on site to find out how bad the situation was. I spoke to Glenn Walsh, who was an ex-Goodbrand guy and the president of Tercon Contractors, and he advised me that they were owed $20 million plus. I had several discussions with Jon Rudolph at Golden Hill Ventures and they were owed a very large number as well, but I don't recall that amount. Al Beaudry from Ledcor advised me they were owed about $6 million and Brian Surerus reported that Surerus Contracting was owed $8 million. Other contractors on site who I spoke to included Ernie Elko at Peter Kiewit, John Cousins at Focus Electric, Duff Plato at McGregor Construction, Vic Bidinski at Vallard Construction, Danny Trainor at

Alligator Installations, Tim Tarant at Nor-West Cladding and John Simkins at Thunderbird Air. We also learned that Findlay Navigation was owed a large sum of money from Royal Oak for the barging work they provided on the lake during spring breakup. This situation was very serious indeed! Another significant complicating factor was that we owed several suppliers large sums of money for work completed on this project including about $700,000 to Lafarge Cement, and other payments to Prince George Steel, Benco Fabricating and Hilfiker Retaining Walls, etc., and unless we were paid regularly by Royal Oak, we had no ability to pay all of these large amounts.

The consensus of this group of contractors, who was really the "who's who" in heavy construction in B.C. at that time, was that Royal Oak was short at least $60 million to pay everybody out for work already completed, plus whatever additional financing they required to complete the project. We now clearly understood that the capital costs of building the mine were well in excess of Royal Oak's original budget, and the other complicating factor was that the market price for gold was now less than the price used in their feasibility analysis. Ultimately, I organized a meeting for all the contractors at the Richmond Inn for December 17, 1997, and most of those listed above attended. Based on our own legal advice from Ray Schachter, we were fully transparent and also invited senior executives from Royal Oak to the meeting to provide them the opportunity to explain how they intended to deal with this situation. Gerald Rockwood, the treasurer, and Ed Szol, the COO, attended on behalf of Royal Oak Mines. What made this entire situation difficult was the fact that all of the general contractors on site could now file a lien on the Kemess property, yet liens would seriously complicate Royal Oak's efforts to seek additional financing to pay the contractors and complete the mine infrastructure. It was a Catch-22 scenario. Generally, what was discussed at the meeting was that the contractors would not likely return to work in January until either all outstanding amounts were paid or there was a clear understanding about how the individual debts would be satisfied. Royal Oak explained they were seeking additional financing and that they were confident it would fall into place very quickly, but as we know, those things can take a lot of time, especially when it looks like the ship is sinking. In the end, we had no alternative but to file a lien on the property in early January for the amounts outstanding to us, and many other contractors did the same in order to protect themselves. By mid-January 1998 there was a total of about $75 million in liens filed by most of the contractors on site.

Negotiations between WVCC and Royal Oak were ongoing for the next 10 months and involved a lot of legal costs, accounting costs and our own time. In the end, I think we were fairly lucky because we were one of the first contractors on the site and most of our work was nearly completed when the cash shortfalls started to occur. Some of the other contractors on site terminated their contracts and ended up in a more protracted legal dispute resolution process. We eventually arranged to meet with Peggy Witte, Ed Szol and Gerald Rockwood in Royal Oak's Seattle office to hammer out a final deal. The setting was on Lake Washington with Peggy's yacht tied up to the dock, and there was a marine feeling to the office, which was eerily out of sync with the world at the mine site. Ray Schachter, our talented business lawyer, and Bill Tough, our comptroller, attended along with Paul and I. After some serious and lengthy discussions and close confrontations, we eventually agreed to accept a partial payment of all outstanding amounts, including all of our substantiated claims for additional costs due to delays and acceleration, with the balance of about $1.1 million paid in equity shares in Royal Oak. While we knew owning these shares was not fully liquid because of the sketchy financial condition of the company, we thought this was the best deal we could possibly make and an early settlement was important before they were declared legally insolvent and their equity shares were worthless. In the end, we leaked all of these shares onto the market over a period of weeks with a net loss of about $300,000, but that amount was peanuts compared to what was originally at risk and what we ended up making in profit on that job. Fortunately, we were able to get all of our suppliers paid and we left the site with a nice profit, but that was a close call for us. Not long after we were completely paid out, Royal Oak became effectively insolvent and was bought for a deeply discounted price by Northgate Explorations, a Bronfman-owned company, which operated the mine very successfully for a number of years.

While all of this was going on over the winter season at Kemess and during the ongoing preparation for arbitrations with H. Fontaine, Telus/

Harbour Electric, Sabre Transport and District of Chilliwack, I separated from Annette, my wife of 20 years, and moved into a house in Crescent Beach that I rented from Nita Manning, Paul's mother. In April, I moved in with my tennis partner Gladys, arranged the purchase of 50% of her house and property from her ex-husband, then began a complete interior and exterior renovation costing about $450,000 over the next year or so. (This might be in my next book!)

Pine Coulee Diversion Facilities - Stavely, Alberta

In April of 1998, WVCC was awarded a project from Alberta Public Works and Supply Services (APWSS) in the amount of $5.7 million for the construction of the Pine Coulee Diversion Facilities. This project entailed, among other things, the construction of a large, reinforced concrete diversion weir across Willow Creek near Stavely, Alberta. Our contract was just one component of an overall plan to dam the river and create water storage and distribution for agriculture in the local area. Bill Campbell was our superintendent for this project along with Dale Lauren as project engineer, Pete Steiner as assistant superintendent, Joe Mather as general foremen, Dennis Case as reinforcing steel foreman, Paul Schlauwitz as batch plant manager and Darren Budzak as surveyor/safety officer. In my opinion, this was an all-star team for the job. Lance Bendiak was the project engineer and Mladen Kovac was the site inspector for Acres International, the designer of the project. We retained McCaw's Drilling and Blasting for the drilling and blasting and Top-Notch Construction for the major earthworks in the contract. Again, in my opinion, we couldn't have found two more experienced and qualified subcontractors to perform this work with us.

Unfortunately, this project got off on the wrong foot early on in construction because the foundation rock below the main weir was not what we believed was represented in the contract documents. It was essential for the concrete weir to be embedded directly against a solid rock face for sliding and overturning resistance; however, our first test blast revealed that the harder sandstone cap rock did not extend to the base of the weir.

Instead, below the caprock we found something called mudstone, which shattered badly with any blasting at all and the dilemma was that the heavily jointed sandstone caprock definitely required drilling and blasting to remove. The other troubling aspect of this mudstone was that it deteriorated very badly when it was exposed to the atmosphere and especially moisture. Coupled with this geological problem, we ended up with one of the rainiest springs and summer seasons on record, so the mudstone literally turned into unstable "mud"! To exacerbate these problems, the designer specified that the diversion weir concrete must be placed in four different sections along its length and in eight separate horizontal lifts in each section to minimize the heat of hydration, thermal expansion and the associated shrinkage cracking in the concrete. As a result, the unstable face of the mudstone was exposed for many weeks or months as this staged work proceeded, causing serious slaking and sloughing while significant quantities of additional concrete were placed up against the downstream native face of the excavation. We ended up with a very loose and ragged face on which to place the concrete, far from what was represented on the drawings or included in our estimate. In some cases, the wall of the excavation sloughed in badly after the reinforcing steel cages were installed, which required us to pull all of the rebar out, re-excavate the trench, clean the rock as best as possible, re-install the rebar and then finally pour the concrete. In other situations, we formed the downstream face of the weir, stripped the formwork and then filled up the void with mass concrete, with limited additional compensation from the owner.

As soon as the initial test blast was performed and the engineer became aware of this change in material, they made some adjustments to the shape of the weir by deepening the toe by 1.5 m and widening it to provide additional sliding resistance, all the while attempting to downplay the impacts from the obvious change in foundation rock that was affecting our work methodology, our progress and our resulting costs. This significant change to our planned work methodologies resulted in a great deal more effort as well as delays to the schedule. Due to the fixed project timetable created to allow the filling of the reservoir in the spring, the owner required us to add more workers and work longer shifts in order to accelerate the work. We also experienced winter working conditions due to the delays and poor results from the acceleration. (Usually, when a serious problem of this nature affects the work, this form of acceleration does very little to improve the schedule and increases costs dramatically.)

All in all, this change in ground conditions turned this job into a complete disaster. The only way we could come out whole was to assemble a formal claim once we were complete and submit it to the owner for all of the additional costs incurred. Unfortunately, the problems we incurred and the accompanying pressure from the owner to speed up and accelerate the work took a terrible toll on the staff's morale, and Bill Campbell, in particular, had a hell of a time dealing with these additional pressures. The project was eventually completed, but we were way over budget and no one had any fun on this job.

The change order request that we eventually prepared was the best change of subsurface conditions claim I have ever seen or heard of, and because the change occurred so early in the progress, we were able to fully document our impacts throughout the work. The claim had clear contractual justification and entitlement due to the fact that the obvious change in subsurface soil conditions resulted in the owner's subsequent change in design. We documented every step of the process, provided contractual notice whenever required by the contract, proved our various efforts of cost and schedule mitigation and assembled a very thorough record of our additional costs. To assist with our claim, we retained Robert Martin, P. Eng., a very well-respected soils engineer from Geo-Engineering, as well as Ted Corcoran, P. Eng., and Alex Chisolm, both renowned blasting experts in Western Canada. All of these professionals were very familiar with the geographical area and the specific geology experienced at the site. They were also extremely supportive of our contractual position and assisted greatly in the preparation of the claim. Mike Demers and Al Morgan, from Revay & Associates in Vancouver, who I had worked with on many occasions, were also very helpful in the preparation and review of the final claim document. The claim was finally submitted in three 4" binders on April 30, 1999, amounting to a change request of $1.2 million. Acres and APWSS completed their internal review of the claim submission on August 26, 1999, and in a telephone call on September 16, 1999, they requested a meeting later that month to discuss the claim. We requested a detailed written response to the claim before the meeting to provide us with a better understanding of the owner's position. I don't think we ever received much of a reply to our claim; nevertheless, we met on September 24, 1999, in APWSS' offices in Edmonton with all of the players attending from APWSS and Acres, including Param Sekhon and John Ruttan, who was the Executive Director of APWSS. Unfortunately, we were not able to make any progress in resolving the claim at that meeting. It was my distinct impression and my belief that many of those attending understood our claim and its potential validity; however, the claim was being shut down from the very top of APWSS. A short while later, the owner responded in a letter denying any responsibility for any delays or additional costs related to the alleged change in subsurface conditions, leaving us with no choice but to exercise our right to binding arbitration under the terms and conditions of the contract. Here we go again, in the midst of another legal battle, despite our sincere efforts to avoid such situations.

A decision I still wonder about to this day, was choosing to hire Ron Kruhlak, an experienced big-firm construction lawyer from Edmonton, rather than retaining Mike Demers, who was from Vancouver, even though we were well aware of Mike's tremendous experience and ability in arbitration and litigation. The sole reason for this decision was that I thought an Alberta lawyer would have a better chance of negotiating a suitable settlement with the Alberta government without having to proceed through the complete arbitration process. Ron and I negotiated a sliding contingency scale with a cap for his fees so that his payment was entirely dependent on our success with the claim however it was resolved. The more we got paid, the more Ron received for fees, so I thought he was fully incentivized. Unfortunately, Ron and I were unable to successfully negotiate a settlement with the owner without proceeding to arbitration.

Over the next several months, we attempted to agree on an acceptable arbitrator as well as a general outline of the process. The contract stipulated that the rules of arbitration would fall under the International Commercial Arbitration Center, which is quite a formal process that includes examination for discovery of all witnesses and the production and exchange of all relevant documents. After APWSS reviewed and rejected a dozen or more potential arbitrators that we proposed to them, they formally submitted three names to us. One of the names was Bob Jenkins. Bob was the arbitrator from Vancouver who had arbitrated the previous dispute on the Bonnington Falls Project and we had been quite happy with his ability. This was a great stroke of luck for us because we knew Bob was very knowledgeable about our industry. More importantly, we were confident Bob knew from our past history that we were honest, straight shooters, and we were also confident he would understand this claim was not just a wild shot in the dark to recover from losses incurred on a project that we screwed up, as some contractors might try. Naturally, we were very pleased to have Bob as our arbitrator and this was formally agreed on April 14, 2000. Dates for the arbitration were arranged for June 19–29, 2000, in Edmonton.

In preparation for the arbitration, we hired an expert witness by the name of Chuck Brawner to assist us. Chuck, who was now semi-retired, had an office in North Vancouver and was a former partner in Golder Brawner, a large geotechnical firm that had offices across Canada. He was a world-renowned geotechnical expert who had literally worked all over the globe in his craft. He had also been a geomechanics professor at UBC and taught Paul Manning when he was an engineering student there. In my opinion, Chuck was the very best in this field, and he had both a wealth of practical experience and technical expertise. He was a straight shooter, and I enjoyed working with Chuck immensely. APWSS retained JDE, a claims consultant, to manage the arbitration for them along with a geology expert named Ray Benson. Both he and Chuck produced expert reports on the issues related to this dispute and you could not find two more opposing opinions! Each party was loaded with ammunition and ready for litigation warfare!

The arbitration finally proceeded in mid-July 2000 and lasted about one week. It was a fairly formal process during which each party's lawyers presented leading evidence, questioned witnesses and cross-examined the opposing witnesses. Unfortunately, the process of binding arbitration under the terms of the contract is strictly confidential, so I am, unfortunately, bound from disclosing any specifics of this process. However, what I can tell you is that we were not pleased with the outcome.

In the end, this entire project was like a kick to the guts. Everyone worked extremely hard on this job, particularly Bill Campbell, but also

Dale Lauren and a lot of our long-term, senior trades including Joe Mather, Pete Steiner, Dennis Case and Paul Schlauwitz. Hourly workers on that job included a lot of our long-time crew such as Wayne Clark, Denis Casavant, Lindsay Berston, Glen Mosley, Ralph Schlauwitz, Mark Wilson, Korey Mather, Bernie Boehme, Robert Kepke, Marcus Steiner, Vern Walker, Gerry McMillan and Gus Gudmundson, along with a number of local people. Although it was certainly not the fault of these workers, I know that the pressure of running over budget and over the allotted schedule affected everyone on that job.

After all the hard work and disappointment from the Pine Coulee project, I was pretty disillusioned and thinking seriously about a career move. It seemed I was on a never-ending treadmill that I had a hard time getting off of. The company's annual volume was all over the map, very high one year with a big project, only to drop off rapidly once that project was completed, leading to a point where we "needed" more work[31]. The problem was, the decreased volume did very little to reduce my own efforts or workload because I was also involved in some degree of estimating and business development when the volumes dropped off. The number of claims and disputes over the last few years had definitely affected my disposition and energy. Then, out of the blue, I was offered a job as bridge manager for Branch Highways, a company in Roanoke, Virginia that did an annual volume of around $1 billion a year in road and bridge construction, and I actually flew down at their expense to learn more about this opportunity. Al Beamer, who had been a good friend of mine since the Coquihalla Highway, was working for Branch Highways, and they really liked their Canadian staff. A lot of things looked really good about this company and their offer, including the company's much bigger financial power, a larger management team to share the load, as well as much lower

31 When I refer to "needing" work, I wish to clarify that WVCC was never a volume-based contractor. We never set targets for volumes of work that we needed mainly due to the widely fluctuating markets we worked. If a new mine or hydro facility came out, we bid them, but if there were no mines or bridges, we didn't go to residential or commercial construction to meet any required volumes. Basically, we bid every single job to make money not to fulfill a target volume. We only lost money on a handful of jobs in 25 years and none of those losses was significant.

housing prices and taxes. However, after careful consideration, I chose not to make the move and carried on with some renewed energy in WVCC. After all, I had become a 30% partner in the company in a previous share restructuring agreement and there would be no point selling my shares at such a low point as post-Pine Coulee.

Churn Creek Bridge - Gang Ranch, BC

After the disappointing Pine Coulee project, we completed another 12 projects worth a total of $6.2 million in the remainder of 1998 and most of 1999. In late 1999, we received another award from M.O.T.H. in the amount of just over $3 million to retrofit the Churn Creek Bridge, which was an old wooden truss suspension bridge over the Fraser River near the Gang Ranch. The bridge was originally built in 1914 as part of the original Caribou Road and was a sister bridge to the suspension bridge over the Fraser River in Lillooet that was built in 1913. Art was again the superintendent, with Dale Lauren as his assistant, and we also had Dennis Case, Denis Casavant, Ralph Schlauwitz, Gus Gudmundson, Joe Gardiner and Mark Wilson as key trades on this job. Ian Sturrock was the project manager for M.O.T.H., and Ron Lyall, who was the bridge construction engineer for the Bridge Branch, became very involved with this bridge due to his particular interest in reviving this old suspension bridge. Buckland & Taylor were the consulting engineers for the Ministry and Roman Cap was our main contact there. As this site was about 100 km by gravel road to either 70 Mile House or Clinton, our workers were generally housed and fed on the Gang Ranch. Some of the key staff lived in the fairly nice ranch manager's house that was vacant at the time, and this was also where I stayed when I was on site. It was kind of cool to walk into the cowboy mess hall for breakfast in the morning.

The existing bridge was composed of one concrete pier on each side of the river with steel towers on top of the piers that suspended the two large diameter steel cables across the river from anchors buried in each approach to the bridge. The project involved the replacement of the wooden through-trusses with galvanized steel trusses and replacement of the timber deck with a galvanized steel open grating. In addition, we had to construct new concrete abutments on steel piles and new anchorages for the suspension cable on each approach. Again, we set up the Fastway batcher near the site and utilized pre-bagged aggregate and pallets of cement that we charged with the cement auger due to the small volume of concrete required. Ralph Schlauwitz and Mark Wilson ran the plant for us on this job. Griffiths Piledriving installed the pipe piles for the abutments and Southwest Contracting installed the soil and rock anchors for the main cable anchorages. Marsh Steel, which was located near our office in Port Kells, supplied all of the structural steel, and we worked very well with Dave Powley on that fairly tricky fabrication project and on many other interesting projects in those days.

One important requirement in the contract was that the bridge had to be open for traffic for a certain number of hours every day, as it was the only access across the river for many miles. That certainly sounded like a challenge!

Churn Creek Bridge Steelwork and Truss Replacement

Now, the interesting thing about bridge trusses is that if you remove one single member of the truss, it will lose all structural capacity and quite possibly collapse, so the methodology for this retrofit was critical. Essentially, we constructed the two new steel trusses piece by piece hanging below the deck and only once those trusses were complete from one end to the other, connected to the suspension cables and carrying the load of the bridge, could we start dismantling the wooden trusses above them. To accomplish this work, we installed a full-width scaffold across the entire bridge span that hung below the existing deck to provide fall protection and worker access. It was supplied and installed by Chinook Scaffold from Prince George, and they were a very professional group to work with. Once the wood trusses were removed, we started removing sections of the deck and installed the new steel floor beams and the steel deck grating. The structural steelwork on this job was like working with a Meccano set, but working with the deck grating was like building a giant puzzle because there were many different shapes and sizes of grating panels. The grating was a special-order material, and we found that there were only two fabricators of this type of steel grating in North America and they were both from Pittsburgh, Pennsylvania.

Churn Creek Steel Decking

We selected IKG to produce the grating, and I worked closely with Tom Mooney, Mike Riley and Tim Pace for weeks on the shop drawings and specifications so they could get fabrication underway. Unfortunately, due to no fault of ours or Buckland & Taylor's, IKG failed to produce the required material to the correct sizing, specifications and schedule, so this became our next nightmare. They acknowledged that they failed to perform (Tim Pace even said in a telephone call their performance was abysmal!) and these problems resulted in delays in shipping, seriously affecting the job's progress. It got so bad that our extra costs due to these delays and fabrication errors exceeded $100k, and while they accepted responsibility on one hand, on the other hand they appeared to be heading towards litigation with us. After many months of negotiation, with the assistance of Chris Armstrong from McLean Armstrong, we were able to avoid litigation in the end. They accepted back charges of $120k from us, while our total claim for delays, legal expenses and head office overheads amounted to $148,000. This nightmare was occurring right in the middle of the Pine Coulee arbitration, so as hard as we were trying to avoid problems like this, they seemed to find us.

Other than the fabrication and delivery problems we incurred with IKG, we enjoyed fantastic cooperation from Ron Lyall, who ended up on site for much of the work, as well as Ian Sturrock and Roman Cap. This was very much welcomed and appreciated due to the number of changes and adjustments that had to be made in the field for this type of old bridge retrofit project. Everyone involved was definitely pulling on the same end of the rope, which was nice for a change. One aspect where Ron was extremely helpful was the final adjustment of all the hanger rods that suspended the new trusses from the main suspension cables to set the new bridge deck grades. As I remember, this required a lot of complex mathematics with the catenary curves of the suspension cables, and Ron loved that stuff. Again, Art and his crew performed beautifully and the job was completed without a hitch, if you ignore the fact that Art was nearly taken to jail at gunpoint by an overzealous fisheries officer who thought a small quantity of spilled concrete on the ground was a threat to the entire Fraser River fishery!

Churn Creek Tower Reinforcement

One funny story that Dale Lauren reminded me of many years later was one of Art's examples of cost-saving, which involved the maintenance of the portable toilets that we had at the site. As this site was a couple of hours away from any reasonably sized town, the maintenance of the porta-potties required a sewage truck equipped with a pump to travel back and forth every week. In order to eliminate these costs, Art devised a system where Dale, as the junior man, would take the "sewage boxes" out of each of the portable toilets each weekend, transport them in his company vehicle to Williams Lake, where he lived, and dispose of the material. Dale didn't sound like he really enjoyed that part of his job, but he was a young, dedicated guy and would do almost anything to make a buck for the company!

Although I cannot claim to have been on this site more than a few times, I was certainly closely involved from the office end of things. It was one of those projects that was a pleasure to be involved in, and even though it wasn't a huge job monetarily, I feel it was one of our company's proudest achievements.

Churn Creek Bridge from Above

Other than the projects in Alberta or the ones way up north that I would generally fly to, many of the remaining projects that we worked on over the years involved a lot of driving as I tried to visit most of the jobs roughly every three to four weeks. The geographical separation of our projects provided me an opportunity to see a lot of Western Canada firsthand. Notwithstanding the often-remote locations where we worked, it became a hobby of mine to drive to or from the various projects by secondary or gravel roads, which I thoroughly enjoyed. After many years travelling in this manner and getting to know various wayward routes, my nickname became "Backwoods". One of my favourite routes was from Spence's Bridge to Ashcroft on the Kirkland Ranch Road that ran on the east side of the Thompson River. This gravel road ran through an old Indian settlement called Pokhaist which featured St. Aidan's Church and beautiful scenery along its entire length. Another great remote drive was from North Bend to Lillooet on the southwest side of the Fraser River. A perennial favourite was the route that bypassed the Coquihalla Highway tolls,[32] followed the

32 The Coquihalla Highway was a tolled highway from 1986 to 2008 and a total of $845 million was collected, which represents the total cost of construction up to 1986.

old KVR through the Coquihalla canyon and re-entered the Coquihalla Highway at the Portia exit. (Unfortunately, this spectacular scenic road has been closed for a number of years due to the Trans Mountain Pipeline, which shares that route.) Other great backroads include the road from the little settlement Pavilion on Highway 99 up to the Gang Ranch via Jesmond and Big Bar Lake, the shortcut from Kingsvale to Aspen Grove on the Kane Valley Road and the old Princeton to Summerland Road on the KVR route. Another great backroad short cut that I drove on a few occasions was the old road from Three Valley Gap to Lumby. There are hundreds of others.

One time, when I drove with Paul, Bill and Blair back from Nelson, we tried to stay on back roads the whole way home. When we got to Keremeos we attempted to drive directly over to Manning Park, and we took hours looking for a route through the high mountains around Cathedral Lake. We eventually gave up, turned around and had to get a motel in Princeton to stay overnight. The thing about back-roading in B.C. is that you will occasionally get lost, but that is half the adventure, especially when it is snowing and/or starting to get dark.

Backwoods Sign Chaylene and Shannon Made for Me at Thalia Lake

Miller Creek Hydroelectric Project - Pemberton, BC

The Miller Creek Hydro Project was another fairly large, privately owned hydroelectric development (IPP) located just outside of Pemberton, B.C. Paul and our estimating staff performed various feasibility and constructability analyses through the fall of 2000 for Knight Piesold Engineering (KPL) and the original owners, David and Nick Andrews, along with Rick Staehli (the group who went on to form Cloudworks). These exercises assisted KPL and the owners to finalize the design and provide a budget for construction. Shortly after this process, EPCOR (Edmonton Power Corp.) entered into serious negotiations with the original owners to purchase this project and eventually closed the deal. Soon after the purchase, we were able to negotiate an alternative-style contract directly with EPCOR for the complete construction of this project in the amount of $28 million, and that contract was executed early in 2001. The contract format was cost reimbursable to a target price, based on the proposed design at that time. The pre-construction meeting was held February 28, 2001, at the Pemberton Hotel, shortly after the formal contract award.

The project entailed two high-elevation concrete water intake structures, one on Miller Creek South at an elevation of 1120 MASL (3675' above sea level) with a low-pressure tributary flow through a 1 km long hard-rock tunnel from the other intake at Miller Creek North. The powerhouse was located down near the Lillooet River floodplain, so the total head was 2500' or 762 m. (The higher the head, the greater the energy potential.) This was the highest head hydroelectric project in North America at that time.

Andy Keller, who was the project manager, was initially our main contact at EPCOR, along with Pat Doyle and Robin Trowsdale, who worked from EPCOR's Edmonton office. Bob Heath came on a short while later as the owner's resident engineer, and one of his main roles was to supervise our construction and to liaise with KPL to finalize the design of the project. Bob became our primary contact point on site once we got fully into construction. Leon Gous was the design manager for KPL, and his team included Sam Mottram and Rob Adams, with Tom Vernon as the primary field engineer and Max Schlagintweit as his field assistant.

Leung Seto was a subcontract engineer who was primarily responsible for the penstock design and Adrian Gygax was another subcontract designer who was primarily responsible for the powerhouse, intake structures and concrete anchor blocks for the penstock. Art was the project manager on this job, along with Dale Lauren as assistant superintendent and Diane Lemieux was a project coordinator. Diane was the latest BCIT grad and came highly recommended from my old classmate, Tom Abbuhl, who was now a department head at BCIT. Area superintendents included Peter Jokinen, who worked on the powerhouse and South Miller intake structure, and Ross Lee, a former Loram and SCI Construction superintendent, was hired to manage the various earthworks operations for us. Trevor Hilton was a great excavator operator who worked for us previously, and he became the penstock general foreman. Dale Lauren also was an area superintendent who looked after the penstock anchor blocks, the pipe bridge and later on, the North Miller intake structure.

One (amusing??) incident that I remember happening during the initial mobilization to the site, that cannot be excluded from this book, occurred in our Langley yard. Apparently, Art and I were working in head office getting things organized to mobilize to the site and he asked Dale ("The Kid") to organize the immediate mobilization of a company-owned office trailer to site. Included in that assignment was arranging to unload the trailer on site as no other equipment was there yet. Dale thought he would send the PC300 excavator up first on a lowbed, then the excavator could off-load the office trailer from the step deck trailer with a couple of wire rope slings. When that plan was presented to Art, he wasn't sure that the excavator had enough height to perform the lift, so Dale decided to hop in the excavator and do a mock-up to see if it would work. When he got the excavator bucket up in the air directly above the centre of gravity of the trailer, he called Art over to show him that he had lots of height and the plan would work. Apparently, when he waved to Art to give him the thumbs up sign, he accidentally brushed the bucket tilt lever and the 3000 lb. digging bucket that was nested into the cleanup bucket for shipment, slid free and plummeted into the trailer, crashed through the roof, then through the floor and eventually destroyed the main axle and the two tires! In a matter of seconds, the Atco-style trailer was totally pulverized and was clearly a complete write-off. All we could do was laugh! I told this story at Dale and Tania's wedding reception several years later in Williams Lake, much to Dale's mortification!

Miller Creek Hydro was truly a greenfield[33] project, so it required the complete construction of 7 km of forest-service type roads and a couple of steel bridges crossing creeks, along with the associated clearing, grubbing, drilling, blasting and the installation of permanent drainage facilities for the roadway, before we could even get to the upper penstock and intake locations. Lizzie Bay Logging, who were located out of Pemberton, performed all of the clearing, grubbing and access road construction on site. They were very experienced in working this steep, rocky terrain, and they performed really well. Norm LeBlanc was my main contact with Lizzie Bay and his brother Doug mostly ran things on site. John Howe was a forestry consultant retained by the owner, who was invaluable in getting the road design and layout completed along with ensuring we had all of the necessary governmental permits as we progressed. The ongoing governmental requirements for this work included permits for tree cutting, use of borrow and waste sites, drainage plans, fisheries approvals for stream crossings, etc. These permits and approvals were essential to maintain on an ongoing basis in order to keep this operation running smoothly. Due to the steep, rocky terrain, the roads were built at grades up to about 22% and involved some sections of sharp switchbacks and gabion rock retaining walls to progress through the toughest sections. Further complicating matters, the road layout had to fit in carefully with the eventual penstock alignment, especially in the steepest area where the road had to switch back and forth. Due to the steepness of the access road, we bought a 6 x 6 powered flat-deck truck from Mainland Machinery, in Lyndon, Washington to haul various materials up the road in virtually (almost) all weather conditions. For maintenance over the winter months, we purchased a used combination snow plow and sanding truck, which was critical to keep the road open as long into the winter as possible. This proved to be a challenge with the heavy snowfall in that area combined with the very steep, narrow access road.

---

33 A "greenfield" project is one located in the wilderness, and in many cases, there are not even any logging or resource roads in place, like in the Miller Creek Project.

Once we established suitable road access around the site, we were able to mobilize our forces to the various work areas such as anchor block locations, the powerhouse and the intakes. Because of the long, steep access road from bottom to top, we decided to install two separate concrete batch plants, one near the top so we could service the higher elevations and another closer to the powerhouse, where a larger quantity of concrete was required. This would greatly reduce hauling time with our concrete mixer trucks, especially uphill. For the lower plant, we set up our complete Fastway plant that was fully equipped for winter and summer concrete. For the upper plant, I recall that we rented a Fastway aggregate batcher from a local contractor in Pemberton and utilized our cement and fly ash auger. This plant was really only intended for spring, summer and fall work because the road would likely become impassable in the depths of winter. All concrete aggregates were purchased from Sabre Rentals in Whistler. Paul and Ralph Schlauwitz ran the batch plants again and performed a variety of other tasks.

The penstock, which was a huge component of work on this job, included about 400 m of low-pressure, 800 mm diameter butt fused HDPE pipe that ran from the downstream tunnel portal to the South Miller intake and 4500 m of welded steel pipe that ran down from South Miller to the powerhouse. The steel pipe ranged between 9.5 mm wall thickness x 914 mm diameter welded bell and spigot at the top end down to 32 mm wall thickness x 1.220 mm diameter butt welded pipe at the powerhouse. One of our first tasks after we were awarded the contract was to purchase all of the steel pipe required for the penstock and that proved to be an exhaustive and time-consuming process. The specified plate material that would be rolled into pipe sections might be in stock somewhere and require rolling, but that was not something you could generally buy in a timely manner locally or maybe even in Canada. On this project, after many weeks of pricing, which included looking at a variety of steel plate options, suppliers and logistical analyses, we awarded some of the pipe manufacture to a supplier in Japan and the remainder to Korea. The due diligence and quality control processes required to purchase $4 million worth of manufactured pipe was very extensive. We ended up purchasing the pipe through North American Pipe (NAP) in Delta, and we retained various testing agencies to perform the required testing and quality control, both in Japan and Korea and again once it landed in Surrey, to ensure we received exactly what was specified by the penstock designers. Steve Siu was a welding engineer who we retained to manage a lot of the engineering and quality assurance. Steve was an extremely knowledgeable welding engineer, and he proved to be a valuable ally. We worked with Steve on many challenging pipe and welding projects over the years. The pipe was eventually delivered by freighter, off-loaded at Fraser Surrey Docks in Surrey and temporarily stored at that facility and at NAP's large storage yard in Delta. From there, it was trucked to the site on highboy trailers with either two or three "sticks" per load as it was required for installation. That was about 120 highboy truck loads!

The penstock route included a number of vertical and horizontal bends to negotiate the terrain and depending on the degree of deflection in those bends, reinforced concrete anchor blocks were required at those locations to resist the additional vertical and horizontal forces. As I recall, we had a total of 47 separate reinforced concrete anchor blocks and depending on the additional forces in each bend, many of those anchor blocks were installed with rock anchors drilled and grouted into the foundation rock. At each bend, the pipe was cut, bevelled and welded to suit the deflection angle required, but depending on those angles, there may have been several mitred cuts to achieve the bend angle. For example, if the bend only required a deflection of three degrees, one mitred cut could be made, but if there was a 20-degree bend, it might require six to seven mitres in the bend. Vic Bilodeau was the welding contractor who we hired to fabricate all of the mitres and perform all of field welding on site. Once the pipe arrived from offshore, Vic and his father, Lionel, along with Jason Hyes-Holgate, got started building the various bends in our yard in Port Kells. Lionel was a top-notch welder who worked with Art on the Burrard Bridge Gas Line Replacement Project in 1997, and that's how we got to know Vic. They clearly remember using our temperamental forklift and lifting and moving the mitres as they made them, with the machine getting stuck and stalling out frequently.

To get started with the bends, Vic had some durable paper templates made up that would lay out the exact required degree of the mitre cuts by wrapping the template around the pipe. Next, guided by the template, they made punch marks on the steel all around the circumference, indicating where to cut the pipe. Vic and Lionel did the initial bevelled cut by hand with an oxy-acetylene torch, then rotated one of the pieces 180 degrees and then lined up the reference punch marks and tack welded the pieces together from the outside. By the way, if the required pipe angle was four degrees, they cut the original mitre at two degrees, then, using the pre-marked top dead centre of the pipe, they rotated the other piece of pipe 180 degrees resulting in a four-degree mitre. Once the pieces were tacked together, they welded from the inside and installed the root pass with a semi-automatic flux core wire feed. Jason then ground from the outside to get to clean metal, resulting in the correct bevels on both sides of the weld, as specified in the weld design. At that point, Vic and Lionel filled the double bevel from the outside with between 12 and 40 passes of semi-automatic flux core wire feed, depending on the wall thickness of the pipe. Overall, I believe there were over 100 mitres in those 47 bends. That was a tremendous amount of welding and an unbelievable amount of grinding for Jason!

EBCO Industries, from Richmond, fabricated the heavy wall bends, reducers and flanged connections in the powerhouse, including connections to the turbine inlet valve and the nozzles to the turbine. They also supplied the heavy-duty manholes for the penstock and bevelled the 48" diameter expansion joint for the pipe bridge in their shop.

The grades of the penstock varied on this rugged mountainside to a maximum of 73%, which I can tell you is extremely steep. The penstock also crossed over one canyon on the route, which required a pipe bridge about 40 m long with a grade of about 45 degrees (100%). Dale Lauren was the superintendent for the installation of the pipe bridge and the construction of the large anchor blocks at each end of the bridge. For the pipe bridge, the sections of pipe were welded together on the ground above the canyon, then insulation and electrical conduits were added on. This combined section was then erected with two rented cranes from GWIL Cranes, a 165-ton hydraulic truck crane at the top of the canyon and a 200-ton hydraulic truck crane at the bottom. To begin with, the lower crane boomed all the way out, right up to the edge of the cliff and connected to the bottom end of the pipe, but did not take any weight at this distance. The upper crane then hoisted the full length out over the canyon to a predetermined point where the lower crane could take 50% of the load and then sat the upper end of the pipe down onto dunnage. At this point, the upper crane disconnected, then picked up the uphill end of the pipe and the two cranes worked in tandem to lower it into the final position on structural steel stirrups that had been accurately surveyed into place at both ends of the pipe bridge. Somerset Engineering provided the engineering for this tricky lift and Frank MacLachlan, the sales manager for GWIL, who we had worked with closely for many years, provided assistance with this pipe bridge erection. Due to the significant bends of the penstock at each end of the pipe bridge, large reinforced concrete anchor blocks were required to carry all loads from the pipe bridge and those were anchored heavily into the surrounding rock with grouted Dywidag anchors. The upper anchor block was particularly challenging to build as it cantilevered well over the face of the rock canyon wall. Dale and his crew of Joe Mather, Wayne Clark, Glen Mosely, Wayne Besler and Dennis Case were instrumental in building these very tricky anchor blocks and the pipe bridge supports. A structural steel support was installed partway down the rock face with grouted anchors supporting the pipe bridge at its midpoint. Southwest Contracting installed all of the anchors for those anchor blocks as well as the centre strut footing.

Miller Creek Hydro - Penstock Pipe Bridge

We started the penstock installation at an anchor block about halfway down the mountain in late October 2001. The planned process for the anchor blocks was to cast the Stage 1 concrete below the pipe and insert adjustable steel saddles so we could accurately set the X, Y, Z alignment of the fabricated bend. Once the bend was accurately located and firmly held in position on the saddles, adjoining lengths of penstock were welded to each side of the bend and fully supported in position. Reinforcing steel and formwork were then installed for the second stage of the anchor block, followed by placement of second stage concrete to fully enclose the bend. We had just completed the first weld to the bend on October 24, 2001, on one of the anchor blocks and I was at the site witnessing that first field weld when I received a telephone call on my cell phone from Tom Vernon, KPL's field engineer. The instructions were to "Stop welding pipe! High-build epoxy coating is required for all penstock pipe." This direction came from Andy Keller, the project manager at EPCOR after months of questions from us about why they planned to install uncoated steel penstock pipe in the ground. Their consistent response had been that any coating was very expensive and simply unnecessary if they allowed enough thickness for corrosion. This new edict resulted in us having to cut the first weld and ship all of the pipe that had been delivered to the site back to Vancouver for coating! We quickly negotiated a price from Seaside Painters to coat all 4.5 km of pipe ASAP in their shop in Port Kells. This was obviously an extra to our contract but it was disheartening to tell Vic that day, "The first weld looks beautiful, but you now have to cut the weld and load the pipe and bend on a truck back to town!" It's better late than never, as they say, but this was a serious curveball in our execution plans.

> Postscript Note: The bad news was that EPCOR only decided to coat the exterior of the pipe at that time and leave the interior uncoated. Only nine years after commissioning the Miller Creek Hydro Project, an investigation took place to determine why the plant was not producing anywhere near its design energy capacity, and it was discovered that the interior of the penstock had suffered from serious oxidation caused by a bacterium that created "turbercles" that protruded up to a ¾" from

> the interior pipe wall causing turbulence and a reduction of flow. The entire interior of the pipe was completely cleaned, power-blasted and coated by others in 2012, and this, along with some other changes, improved the plant's output dramatically.

The entire penstock was fully buried other than the pipe bridge section. The purpose of burying the pipe was threefold: first, the compacted soil around the pipe partially restrained the lateral and longitudinal forces in the pipe when in use; second, it helped eliminate internal pipe stress created by thermal forces from the sun and the exterior environment that exacerbate the hoop stresses in the pipe caused by the internal pressure; third, it protected the pipe from vandalism. The development of the penstock right of way involved a total of 80,000 m^3 of mass common excavation with another 40,000 m^3 of trench common excavation and 10,000 m^3 of rock excavation. Trench bedding and pipe backfill included the placement of 20,000 m^3 of processed granular fill that was compacted to at least 95% standard proctor and closely monitored, as the penstock design relied heavily on the physical restraint of the pipe from backfill. All bedding material was processed on site at a large borrow area part way up the access road, but it was very silty in nature and we had to add additional rocky material from a source at the bottom of the hill, utilizing both a jaw crusher and a cone crusher to eventually produce the specified material. Sabre Transport, out of Whistler, did all of the crushing and screening, but due to the silty, gap-graded material on site, it was a constant battle to produce the specified material and it ended up being a very expensive operation.

Once some of the pipe coating was completed and cured in the shop, we got restarted on the installation of the penstock. We generally used Cat 583 side booms, and when necessary, we assisted with a variety of excavators that were working along the trench. We created "weld bells" (basically a gap in the bedding) as well as timber dunnage below the pipe to provide suitable access for the welders to work all around the pipe at each weld location. The field welding of the penstock was much easier than the mitre welding in the yard as all of the pipe came with machine-prepared bevels at each end, so all of the welding was completed from the outside, then the route pass could be ground out and rewelded with one pass from the interior of the pipe. As I recall, the removal of the root pass and rewelding from the inside was a requirement of Section VIII of the Boiler and Pressure Vessel Code that was applicable to the design of this penstock, although not typically required on gas and oil pipeline jobs. Again, Vic, along with Glen McDonald and Erin Swan, performed all of the welding, with Lionel basically being the welder foreman and keeping everything organized to keep the welders busy.

Stasuk Testing performed all of the site weld testing and Nigel Lowe was their man in the field. The penstock design specified that every weld required 100% visual and ultrasonic testing. Now, I won't pretend to be knowledgeable on interpreting ultrasonic weld test results, but I think there is some degree of personal interpretation involved, and I think Nigel found some possible minute flaws in the welding once or twice that Vic and his crew took great exception to, creating some friction on the job. Maybe Nigel was tired of approving every square inch of every weld, so he needed some excitement, who knows, but I think Nigel worked hard to perform his job. I know that on the several miles of steel penstocks Vic and his crew welded for us on several projects, there might have been only one or two very minor repairs that were ever required.

Once a stretch of penstock was welded, tested and the exterior coating repairs completed, we used several excavators along with the side booms spaced out along the length to lift the welded pipe section with nylon slings, remove the dunnage and place the pipe in the trench on sandbags set accurately for elevation and grade. On those days, with all that equipment working in unison, we really looked like pipeliners! On some of the steeper areas alongside the access road switchbacks, we constructed road spurs to provide access for hydraulic cranes in order to install the individual pipe sections in that area, which were about ½" wall thickness and weighed about 10,000 lb. Once the penstock was in place, separate crews backfilled and compacted the material that we processed in the pit located about halfway down the mountain. Thirty-ton Volvo trucks transported all of the excavated material on site, as well as the processed backfill material all along the penstock, so it was common to have two-way hauling going on all over the mountain. At one point where the penstock crossed underneath the access road, we had to install the pipe right up to the road, create a short detour over the pipe, then carry on below the road. This was logistically tricky to do with huge volumes of excavation on each step of this very steep area (55% pipe grade and 20% road grade) because we had to maintain continual road access. Thankfully, Gord Trottier came to the rescue that day, running one of the 400 excavators and occasionally jumping in the D8 Cat dozer and ramming away until it was done, and we completed the switchover in no time. I was on site witnessing Gord's work that day and it was just as impressive as what I had seen him do in North Bend six years earlier.

Miller Creek Hydro - Some Tricky Steep Sections of Penstock

The steepest section of penstock was down near the powerhouse, so the wall thickness was around 1" at a grade of 73%. The installation of this pipe required extensive research, and I spoke to many old timers with serious pipeline experience throughout Western Canada to look at various possible methods. In the end, we used two side booms per length of pipe that were connected to a large winch anchored at the top of that slope. After exhaustive research, literally around North America, I found the perfect winch for this operation from Rasmussen Equipment, which was located in Belle Chasse, Louisiana! These people had winches and knew exactly what we were looking for. This Skagit winch was equipped with two independently operating drums and controlled by Vicon converters, which provided us with feather-light adjustments and huge pulling power.

Each pipe on this slope weighed over 20,000 lb., and the two side booms weighed a total of 200,000 lb., so we needed the horsepower. The winch itself weighed 32,000 lb. and was equipped with 1-1/2" 6 x 26 diameter wire rope that had a breaking strength of 231,000 lb., so we could almost pick up the loaded side booms vertically! The winch was shipped to us via Rasmussen's yard in Seattle Washington, and it arrived on a lowbed around June 30, 2002. I happened to be on site that day and heard over the truck radio that it had arrived on site. Knowing the size and weight of this winch and the trouble I had finding this particular unit, I got on the truck radio and told the pipe handling crew to be very careful unloading the winch, as it was worth $250,000 and had been shipped all the way from Louisiana for us. Do I need to tell you what happened to the winch? Apparently, my call jinxed the operation. The people in charge of unloading made a slight mistake in rigging and the winch rolled onto the ground when they picked it up with the crane! Well, to make a long story short, we loaded it back onto the lowbed, trucked it directly to Rasmussen's Seattle yard, had them complete the repairs and we had it back on site within one week. Once we got back into action on this hill, we mounted the winch onto a timber bed on horizontal ground just over the top of the steep slope, then anchored the winch back with many wraps of wire rope to buried logs acting as deadmen, similar to what we had done at the Carnes Creek Bridge and again at the Alexander Creek Bridge. Again, Somerset Engineering provided this engineering for us. Initially, the winch was used to suspend an excavator on the hill when excavating the pipe trench, then we hooked up the pair of side booms to install the individual pipe lengths. The entire operation worked flawlessly with one drum pulling the side booms and the second drum pulling the pipe that was suspended by the side booms. The two independently operating drums on the winch provided us with the necessary alignment of the pipe to allow the welders to achieve a proper fit up without too much trouble. Once the welding and coating repairs were completed on this steep section, we placed cement treated backfill (CTB) around and over the pipe with excavators winched up and down, as it would have been virtually impossible to backfill conventionally on this steep slope. Again, like the Soo River project, the CTB performed beautifully on this steep slope with minimal placing and compaction effort.

Miller Creek Hydro - Penstock Installation at 73% Grade

Pacific Blasting was awarded a subcontract to construct the 1 km hard-rock tunnel from North Miller to South Miller. Ron Elliott was the manager at Pacific Blasting and my contact for this project. Phil Read, who was our tunnel engineer at Soo River, was the project manager for Pacific Blasting. Phil ran the work at site for the first year and Mike Petrina took over the second year, with Lewis Clarke as assistant project manager throughout the project. The size, length, grade and rock conditions were very similar to the tunnel that BAT Construction drilled for us on the Soo River Project, and the selected tunnelling method was pretty similar. Once the access road was completed to the South Miller intake, Pacific mobilized to the site and established a portal in order to commence tunnelling. The tunnel was designed to be 9' x 9' square and they drilled and blasted 8'–9'rounds with an electric single boom jumbo but used a "longtom" as a backup machine. The "longtom" was basically a scoop tram with multiple jacklegs mounted to the front of the machine. All of the drill muck was eventually removed to the portal area with diesel-powered scoop trams and was stockpiled locally. Pacific also drilled and blasted re-muck bays, which were short off-shoots in the first half of the main tunnel where they could quickly move the muck from the face of any round in order to get the drill going on the face as quickly as possible, then the muck could be removed from the re-muck bays concurrently with the drilling process. Pacific generally blasted one complete round per shift, and they worked two shifts per day. They encountered some poor-quality rock and some badly fractured zones of rock throughout the length of the tunnel where they were required to install various types of tunnel support under the terms of their subcontract. This ranged from Class I to Class V tunnel support, which included combinations of rock bolts, shotcrete and steel sets. They were paid at the contract unit rate for whatever support was required for safety and had been approved by the engineer.

Once Lizzie Bay completed the access road to the North Miller intake, we had to find the best location for the tunnel portal, but there was a large quantity of soil and talus material that needed to be excavated before we could actually develop a face to start tunnelling.[34] We reached what we thought was a reasonable face of rock, only to have Pacific quickly determine that the rock was very weak, requiring soft tunnelling techniques until competent rock was encountered. For this section, they employed some techniques similar to the New Austrian Tunnelling Method, which was essentially a combination of sequentially shotcreting, pressure grouting and rock bolting the circumference of the tunnel and using steel sets where required. This was a costly and fairly slow process, but once they reached competent rock, they drilled and blasted the remainder of the tunnel conventionally from this heading. Overall, Pacific was on site approximately eight to nine months to complete this work.

The South Miller intake was a reinforced concrete structure that crossed the river at a strategic location to divert water flows into the penstock through a headgate. Peter Jokinen was the area superintendent for this intake structure along with his right-hand man, Jari Mackinen. They built the structure in two sections, diverting the low water flows around the work area during the cooler fall and winter months before access became impossible due to winter road conditions. Once the structure was completed, the water level of the head pond upstream of the structure was controlled by a large structural steel crest gate that operated hydraulically.

---

34 Even though we had already started tunnelling from the south portal, we could always adjust the alignment to meet the desired north portal by tunnel survey.

The crest gate that we installed was fabricated by Steel-Fab Inc. out of Fitchburg, Massachusetts, and my contact at Steel-Fab was Louis Bartolini, along with a local sales representative, Scott Salzer, who operated out of Seattle, Washington. This was a complicated specialty item designed to function remotely in freezing cold weather and built to accommodate a certain amount of storm events with debris flows and siltation, so we worked together for months in the design, fabrication and installation stages. We experienced some design and fabrication problems with this gate, making the process a lengthy one, but it was completed in time to start the plant commissioning.

The North Miller intake was a much smaller, simpler concrete structure with a rock fill dam adjacent to it that acted as a stationary weir directing a certain quantity of water through a pipe to the tunnel. A minimum riparian flow was also maintained consistently for North Miller Creek with another outflow pipe through the weir. At one point in the fall, during construction of this intake, there was a serious rainstorm in the Pemberton area that melted recent snow accumulations in the mountains and caused extreme flows and flash flooding in the rivers throughout the area. Miller Creek was not spared and we experienced flooding at North Miller and around the powerhouse. Luckily, the intake structure at South Miller was completed and held very well without any significant damage. Dale Lauren was in charge of the North Miller intake and could see the water rising quickly as it threatened to wash out the entire rock berm and the partial concrete structure that was already constructed. Luckily, we had a 300-sized excavator at that location, and Doug Leblanc from Lizzie Bay was nearby, so they just started bailing this rocky material onto the berm to strengthen it. Somehow, they were able to bail enough material faster than the material eroded from the rushing water and this action saved the intake. The powerhouse was not damaged but the rushing waters eroded some ground adjacent to the building and created a mess that had to be cleaned up. We were quite fortunate that we suffered such little damage over the entire project, whereas Rutherford Creek Hydro, which was another IPP being built by Peter Kiewit in another valley nearby, suffered millions of dollars in damage with entire penstocks and concrete structures washed out.

The powerhouse was a fairly large cast-in-place concrete structure requiring about 1100 m^3 of concrete that was formed and placed by Peter Jokinen and his crew. Colony Management supplied and installed a pre-engineered building on our foundations, which housed a single 30-mw vertical Pelton turbine and an auxiliary 3-mw horizontal Pelton turbine. Art Gairns was the project manager for Colony and a very cooperative and knowledgeable guy to work with. The main turbine supplied by Alstom was equipped with a power-former generator, which avoided the necessity of a separate step-down transformer. Lethbridge Welding worked with Alstom on the installation of the turbine generator equipment and Westpark Electric installed all of the generator and switchgear equipment and wiring as a subcontractor to Alstom. In addition to that work, Westpark was a subcontractor to us for the balance of plant electrical scope of work as well as the installation of the electrical conductor and fiber-optic line to the South Miller intake for flow control measurement and remote crest gate operation. As usual, Westpark performed its work beautifully without any issues. Westpark had evolved since its inception in the mid-90's, and Steve Jones was a key foreman for the company since the beginning and stayed on with them for many years to come. Bill's son D'Arcy Soutar became a key piece of Westpark as he generally worked in the office in Chilliwack while Bill spent most of his time in the field where he enjoyed the work. D'Arcy was definitely a chip off the old block, and he was carrying on the traditions of innovative work and reliability with great service to the owner or general contractor. D'Arcy eventually became president of Westpark, which allowed Bill to do what he enjoyed most—working on projects in the field.

Some of the regular key tradespeople on this project included Lloyd McHarg, who ran the 60-ton crane; Rick Johns, who ran the 28-ton crane, the 6 x 6 truck and the driving mixer trucks; and Wayne Clark, Paul Schlauwitz, Ralph Schlauwitz, Nick Maskery, Glen Mosely, Zane Lush, Wayne Hibbert and Lindsay Berston. Most of our regular crew were living in the Whistler-Pemberton area, but I was reminded during the writing of this book that Wayne Clark, a single guy who loved the solitude of the outdoors, decided to camp in a tent. (By the way, all out-of-town people, including Wayne, received a healthy living allowance that enabled them to have good, clean, local accommodations.) This camping went on right into the winter until one day it snowed so much while he was at work that the tent collapsed, so he moved into a motel that night!

We had acquired much more equipment over the previous few years and had added quite a bit more to the fleet for Miller Creek through either direct purchase or lease to purchase arrangements. We had about seven to eight excavators working on this project, including a Hitachi 230, several 300-sized machines and about three 400-sized machines, plus about seven 25-ton and 30-ton Volvos, a D8 Cat, a D6 Cat, several loaders, two IT-18 tool carriers and two or three cranes at any one time. Gerry Nelson, another North Delta High School product (brother of my first sweetheart Jacquie) was our mobile mechanic, and with so much equipment operating on this rugged site, we kept him very busy. He was a great asset to have on site, and he just kept plugging away to keep everything operating and "shiny side up" as best he could. Jim Mutter, who was a shop superintendent for Goodbrand in the 1980's, was now working in our head office as equipment manager, and he was invaluable to us for purchasing and disposing of equipment, taking care of warranty items and assisting all of the jobsites with their equipment purchasing and maintenance issues.

The Miller Creek Hydro project was a huge amount of work and in order to succeed, it took a significant effort by our head office and field staff to manage this work over three different construction seasons. These types of projects were extremely difficult because they had so many unknowns going into the work and millions of important details to be resolved. Success in these projects was all about the details. In addition, the complex structure of the design team with KPL leading the way, Gygax providing structural engineering and Leung Seto providing penstock design also created additional work and some stress. Granted, these types of projects are extremely onerous for designers as well because it is a "design on the fly" type of program as the site progressively opens up. Original design is based on LIDAR survey information that is collected with an aerial flyover, which is great to get started but eventually everything needs to be connected to an accurate ground survey, which can change things. Similarly, ground and subsurface conditions are generally unknown during the original design stages; however, as the site opens up and ground conditions become known, these designs need to be adapted accordingly and final designs provided quickly to the contractor to minimize delays in getting the field construction underway. Luckily for us, Tom Vernon was involved in the field end of the engineering process from the beginning of this project and he proved to be invaluable for maintaining practicality and cost-efficient designs as best as he could in his position. He was also a fair guy to work with and he could make independent decisions, which was absolutely necessary to keep the momentum on this project. Bob Heath was also an excellent representative of the owner. He was very practical minded and understood what our work involved. Without Bob and Tom, this project could have easily gone off the rails.

One other taxing aspect of this project for WVCC was dealing with EPCOR and their corporate culture. Typically, they were a large thermal energy provider from Alberta, where they burned coal for energy. They were on a very steep (pardon the pun) learning curve with building a relatively small hydroelectric project on such steep, difficult terrain, not to mention dealing with all of the provincial regulatory bodies that were involved in B.C. for this type of project. One example of this additional effort required by us on this project was managing the constantly changing contract format and the evolving budgetary pressures exerted by EPCOR. When we initially started, we provided a cost-reimbursable-to-a-target-price budget based on our mutual understanding of the scope of work and the baseline set of drawings we were provided, and that contract was executed. As final designs evolved over time and changes were made by the design team through the process, our target price was adjusted for 'material changes' in accordance with the terms of that contract. Those increases in cost were becoming concerning to EPCOR management, including Robin Trowsdale, who was one of my counterparts in Edmonton. However, the design team was not under our control, so we could only price the changes when we received them in the Issued for Construction drawings, and these changes resulted in price escalation. In late January 2002, EPCOR wanted to close off the target price contract and just pay us for the remainder of the work on a cost-reimbursable basis, which we accepted. This went on for part of the next season with a great deal more paperwork and transparent reporting, as is required on this type of contract, including labour time sheets, equipment time cards and copies of all invoices. Later on, Robin approached me in early December 2002 and requested a final lump sum to complete all of the remaining work as he explained to me that he could not keep going to the Board of Directors asking for more budget every few months. I advised Robin that a lump sum price would result in increased risk to WVCC for any possible quantity overruns and continued design changes. He clearly understood, but he was agreeable to a lump sum price that included our increased risk exposure. So, we re-estimated the remaining work, including an allowance for continuing risk, for the final time and entered into a final lump sum contract to complete the work. In the end, the Miller Creek Hydro project was a very successful job for us, but again, it was a very taxing job on the staff. Everyone had to work extremely hard to make it a success. Any number of things can and do go wrong with this type of project, and it takes collaboration between a very experienced group of professionals to ensure they are successful. A great many of the IPPs in B.C. have turned into financial and legal disasters for a variety of reasons, and many contractors avoid this sector of work entirely or regret getting involved in it for this reason.

Hyland River Bridge Deck Replacement - Watson Lake, Yukon

The Hyland River Bridge Deck Replacement was a contract with Public Works and Government Services Canada for a retrofit of an existing bridge deck on the Alaska Highway near Watson Lake, Yukon. The bridge was a four-span steel truss superstructure originally built in 1943 as part of the construction of the Alcan (Alaska) Highway for the war effort undertaken by the US Public Roads Administration and US Army of Engineers to create a highway link between the continental US and Alaska. In February 2004, we received the award for the bridge deck retrofit in the amount of $4.2 million and we attended a preconstruction meeting on March 9 of that year. Dale Lauren bid that job and we ended up being low bidder, so he ran the project at the ripe age of 26! Dale had been mostly working under other superintendents or project managers since 1996, but he always showed a great deal of experience and common sense, so this was a great opportunity for him to run a project on his own. Dale recently recounted to me that he left sunny Vancouver on April 15, 2004, with all of the cherry blossoms bursting in vivid colour and arrived in Watson Lake two days later, having to climb over an 8'-high snow bank to get into his rental home! His wife, Tania, was clearly distraught with this dramatic change in climate and geographical location, but Dale consoled her that there would be no mosquitos for at least the next month!

The scope of work included completely removing the concrete-filled steel deck grating; strengthening the existing trusses; installing a new, wider, steel deck grating and casting a new concrete bridge deck and parapets. The work also included structural changes to the abutment bridge seats and the pier caps to accommodate the wider deck and the revised deck joints. Again, this project required one lane to be operational for traffic at all times, so we were only able to work on one lane at a time. The schedule was to complete both lanes in a single season so neither we nor the travelling public would be negatively affected by winter weather. Stan Reimer and Don Stanek were representatives from Public Works Canada (PWC) who operated out of the Fort Nelson and Edmonton offices. Ramon Callattung was the project manager for PWC who supervised the work on site. Delcan Engineering was the design consultant and Hugh Hawk was the design engineer I was in close contact with.

As soon as this contract was awarded, I finalized subcontracts for the supply of both structural steel and steel bridge deck grating as that material would be required on site very quickly once we got the demolition underway. The structural steel supply contract was awarded to Marsh Steel in Surrey, B.C. and included new steel girders, diaphragms, stringers and truss-stiffening members. My contact at Marsh was again Dave Powley, and Ron Marsh was the shop superintendent. Because the new steelwork needed to match the existing riveted structure, we performed comprehensive as-built drawings of the existing steelwork so everything matched up during the erection. Fortunately, we had worked with Marsh Steel before, and they were located very close to our office in Port Kells, so it was very convenient to work with them in person on the various details for shop drawings. The steel deck grating supply contract was awarded to LB Foster who were located in Pittsburgh, Pennsylvania. Oddly enough, I was now dealing with both Mike Pace and Mike Riley at LB Foster, who were my main contacts at IKG for the steel grating for the Churn Creek Bridge about four years earlier. If you recall, we had some serious and avoidable fabrication and delivery problems on that job, so I decided to be proactive and fly to Pittsburgh to work out all of the details so we could avoid those problems. That trip on March 7, 2004, was well worthwhile, and I pledged to make those type of trips more often with major suppliers. The expense and time required for a trip like that is nothing compared to the expense and time of discovering fabrication and fit-up problems in the field. LB Foster performed beautifully—their shop drawings were completed quickly and both fabrication and delivery were excellent.

After finishing my meeting at LB Foster's shop that day, I ended up with several hours to wait for my return flight to Vancouver, so I decided to go for a drive. After thinking for a few minutes, I decided to drive to Latrobe, Pennsylvania, the home town of Arnold Palmer, which was only about one hour's drive away. So, off I went and found the Latrobe Golf Course in no time at all. I bought a couple of golf shirts, picked up a few score cards and some other souvenirs, took some pictures and headed back to the airport. Visiting the golf course was very interesting because I saw Arnold's pro shop, his souvenir building and his club-making shop. It was quite an experience, particularly because I am a huge fan of Arnold's and had started collecting Arnold Palmer memorabilia some years prior.[35] My visit that afternoon certainly added a few things to that collection.

I must also add that I loved the small city of Pittsburgh. It was nothing like I had imagined steel towns to be like, with steel mills and pollution. Pittsburgh was relatively uncrowded, clean and modern, yet still had some beautiful historical buildings. Although there were no longer any steel mills in the city itself, I saw some big steel mills not that far away on the Ohio River when I was flying out of town.

35 Stan Stewardson, my basketball coach from high school, was an avid sports collector who got me started collecting Arnold Palmer stuff by bringing a little gift every time he came for dinner at my house. Over the years, I received an autographed picture of Arnold, a Palmer shoe bag, video games, posters, coasters, golf instruction books and all kinds of collectibles.

The general methodology for the deck retrofit was quite similar to the Fraser Hope Bridge, where we sawcut the deck in sections, then used a boom truck from the existing deck and girders to remove the deck components and install the new structural steel. However, due to the spacing of the girders, we needed to find a suitable boom truck with the required lifting capacity and outriggers that would work with that girder spacing. I eventually located a potentially suitable boom truck from Myshak Crane Rentals in Edmonton. It was a 21-ton Peterbuilt boom truck and we retained Somerset Engineering to certify this crane to perform what we required. They performed a test lift procedure in Myshak's yard and had it stamped by a P.Eng., confirming the crane could do our required lifts with both rear outriggers down and only a single front outrigger in place. All lifts on the bridge were required to follow this exacting lift procedure very carefully. The crane was picked up on April 22, 2004, by Eldon Kohler, an experienced crane operator from Edmonton recommended by Myshak. Eldon drove the boom truck to the site where they immediately began the work.

Dale's crew consisted of a lot of our long-time regular employees, including Joe Mather as general foreman, Dennis Case, Wayne Clark, and Glen Mosley. Joe and Dale recently reminded me that one local Indigenous fellow they hired on the jobsite literally travelled by canoe to the job every day, as he apparently lived in a tent somewhere upriver! Speaking of tenting, our long-term employee, Wayne Clark also enjoyed personal isolation and remote camping as he did at Miller Creek, and he set up a tent very close to the bridge for much of the schedule. Apparently, he was visited by bears on more than one occasion and added an electric fence around his tent!

The complications of this deck retrofit involved the coordination of many details between the structural steel, the steel deck grating, and the deck expansion joints. It was required that we locate every Nelson stud and splice plate bolt on the top flanges of the stringers, as well as determine the splice plate thickness for the top flanges to suit the spacing and depth of the grating bars. Every deck panel was also equipped with levelling bolts to set the grade of the deck, depending on stringer cambers and haunch heights. It was like an adult Meccano set on steroids. Hugh Hawk was excellent to work with, resolving all of these details quickly, using practical common sense in all cases. Hugh and the PWC staff were also very helpful with assisting us to locate suitable concrete aggregates, which we eventually purchased from Blue Canyon Concrete located in Fort Nelson, who I had worked with back in 1983 on the Fort Nelson River Bridge. The concrete we batched for that job was also a high-performance mix design (similar to the Fraser Hope Bridge) that included the use of fly ash, silica fume and a corrosion inhibitor both on the substructure modifications and the deck. Precast parapets were manufactured by Kemp Concrete from Kamloops, and we erected them on site once the deck was completed.

One unfortunate incident occurred early in July when Eldon swung the crane around to pick up a very light item, totally forgetting that the crane was not to make any lift except between the rear outriggers. The crane tipped over and the boom crossed over the open lane of the Alaska Highway, closing it down until Dale and the crew could get equipment to the site to right the crane. Luckily, no one fell off the bridge or was injured and the damage to the crane was fairly minimal, but Dale fired Eldon on the spot. The remainder of the bridge was completed without incident and Dale and his crew departed on October 15, 2004, with PWC very pleased with the workmanship and cooperation of our staff.

Dale had become a very good superintendent over the past number of years, and the Hyland River Bridge was a great example of his ability to manage a complex project in a remote area. These types of projects were not easy because we didn't have a hardware store or a local delivery truck just around the corner, and we had to think ahead if we needed something and order it well in advance, including additional parts and equipment-servicing components. Dale had clearly become one of our "A" superintendents, and on one day I was on the site, I asked him to go for a walk with me to discuss his great future with the company. I wanted Dale to consider buying some company shares so that he could participate in our profits. However, Dale announced to me that he had been making other plans to go into his family business with his brother. As shocking and disappointing this news was, I supported Dale's move, and he has been running a very successful construction business in Williams Lake ever since. Still, it was a tough blow to the company because Dale had become so valuable to us.

In September of 2004, I was able to sneak away to Europe with my second wife, Gladys, for another well-deserved getaway. We were married in the summer of 2001 and had been planning this trip for quite a while. This trip took us to England, Scotland, France, Italy and Greece, and the most enjoyable points were Santorini, Greece as well as Cinque Terre and Tuscany, Italy. This was a fantastic trip, and it was nice to be free from the grind for this six-week period.

Another development in my personal life in 2004 was my purchase of a very well-preserved Honda Goldwing motorcycle. I had kept the Honda 250XL since I bought it new in 1979 while working in Penticton, but it was stolen in 2000, and I missed the ability to go for an occasional ride like I had for many years. I knew that someday I would get into a nice Goldwing because they were the Cadillac of motorcycles with all kinds of comfort, power and accessories. The day I brought the Goldwing home Gladys said to me, "Have fun, I am never going on that!" However, soon after buying this bike, we discovered that two couples who we played tennis with at Hazelmere Tennis Club also owned larger cruiser bikes and they were interested in going on trips with us. This began an excellent hobby for us that would last for many years to come. Our first long-weekend road trip with our friends Don and Arden Comber and Earl and Cathy Karam was up the Sunshine Coast to Powell River, across on the ferry to Comox and up the Island Highway to Campbell River for the first night. The next day, we travelled north to Port Hardy and returned back to Campbell River, where we stayed at the Discovery Inn, right on the water.[36] The following day was over the windy, hilly road to Tofino for the night, returning home the next day via Swartz Bay, near Victoria. The trip was a complete blast! As all biker groups have names, we decided on a suitable name while we stopped

---

36 That fateful night in Campbell River was the night when I was having so much fun on that ride with the group that I signed a dinner napkin stating I would take them on an all-expenses-paid boat trip to Desolation Sound on a boat I would charter and skipper! Well, no one let me forget that promise and the very next summer we cruised from Richmond up the coast and all over Desolation Sound on a nice 36' Bayliner. That was truly a great trip. Over the years, I have chartered many power boats up to 43' in length and cruised through the Gulf Islands, the San Juan Islands or Desolation Sound, which is always a great week in August!

for lunch in Port Hardy. The name was based on the fact that the women had to inspect several cafés and pubs before any of them were deemed appropriately clean and modern enough for lunch. The name we picked was the FFMC, which stood for the "Fussy Fuckers Motorcycle Club". We went on many trips together, including the Duffey Lake loop, back and forth up the coast to Lund several times, around the west Kootenays a few times, up through Tumbler Ridge and Fort St. John once, through the Rockies to Edmonton a few times and once down Highway 22 from Drayton Valley to Pincher Creek, then back home through the Crowsnest Pass. We also did a great trip down the Oregon Coast, inland over the passes through Oregon and Sun Valley, Idaho; through Missoula, Montana and back through Idaho and B.C. I am now on my third Goldwing and still love touring around to this day. Perhaps my most scenic motorcycle trip so far was three weeks in southern Utah, western Colorado and northwest New Mexico. This was literally day after day of beautiful scenic red rock canyons including Zion National Park, Bryce Canyon, Arches National Park, Telluride, Ouray, Durango, Albuquerque, Santa Fe and Taos, New Mexico. I hope to be able to do that entire trip again in the near future.

In later years, I purchased a second motorcycle, a 1200 cc Yamaha Super Tenere, which is an on-/off-road bike that I rode four times from San Diego to Cabo San Lucas in the winter months with a group of serious motorcycle enthusiasts. That is a fantastic trip with beautiful weather, windy scenic roads, mountain passes and great experiences though old Mexico. The trip takes six days to get to Cabo, then six days relaxing at the hotel pool, then miraculously we make it back to San Diego in four or five days.

In 2005, my brother Bill, who had been our comptroller for the last 14 years and a heavy smoker, was diagnosed with oral cancer. The doctors initially thought that the condition was operable and he would recover, but once they operated, they found the cancer had spread much more than they originally thought. He was only given about three months to live and there was no treatment to cure or curb his disease. He immediately took leave and convalesced at home in his condo in White Rock. We then hired Lesley Huygen, who was referred by BDO Dunwoody, our company accountants, to replace Bill and she took over his duties. Bill tried to enjoy the remainder of his days as best as he could listening to his collection of music, reading books more voraciously than he ever had and watching old movies and television programs from the 50's that he pirated off the internet. Bill outlived the doctor's prediction by about one year, but sadly, he died on November 6, 2006, which was my father's 82nd birthday. Fortunately, he only spent one night in the hospital, and he was able to live on his own right up until the end.

Harrison River Bridge - Harrison Mills, BC

The Harrison River Bridge was another very interesting bridge retrofit project that was awarded to us by M.O.T.H. on April 20, 2005, in the amount of $6,700,000. The existing highway bridge located on Highway 7 near Harrison Mills was built in 1956 and the superstructure was in very poor condition. The bridge had two stationary spans and a centre swing span that allowed marine traffic to pass through, providing access between the Fraser River and Harrison Lake. The scope of work in this contract included the complete removal and replacement of the concrete-filled steel deck grating and the exterior girders on both stationary spans. There were also structural changes to each abutment to accommodate the wider traffic lanes and new bridge deck expansion joints. In addition, it was required we strengthen and repaint the existing structural steel at the pier and abutment locations, replace all deck expansion joints and remove all steel curbs and railings and replace them with reinforced concrete parapets on the two stationary spans. On the swing span, the work was generally limited to the retrofit of the deck joints and replacement of the existing guard rails with new steel parapets. A special 6-mm, skid-resistant wear surface was required on the entire refinished deck surface. Again, like the other bridge retrofit projects we had completed, it was necessary to have traffic operational in one lane at all times, and that always made things very interesting.

Val Fabick (then later, Matt Choquette) was the project manager for the Ministry with Tom Bayntun as the project supervisor on site. Buckland & Taylor was again the consultant with Sergiu Aroneanu and Andrew Griezic as the design engineers. Art was the superintendent for this job along with Diane Lemieux as assistant superintendent. Kyle Jones, the son of John

Jones, who was an old friend of mine from the Goodbrand days, came in partway through the job (after graduating from BCIT) to assist Art and Diane manage the project.

The proposed method of construction and staging suggested by the consultant in the contract documents was to complete most of the work from floating barge-mounted equipment located below the bridge. Because this area around the bridge was a shallow "tidal"[37] waterway and an important environmentally sensitive fishery, we believed that the prescribed method would impact the marine environment significantly and we would be limited to work only at higher water levels, which could greatly limit the available days with this method. Specifically, regarding the environmental concerns, working with marine-based equipment would involve a significant amount of work by tugboats moving the barges around in the shallow area, resulting in environmental impacts from prop-wash stirring up the muddy bottom and from the setting and removal of the steel spuds into the mud that would be used to secure the barges in position. With all that in mind, Paul and I decided to make a trip out to the site on a Sunday morning prior to bidding the project to brainstorm possible alternative methods to perform the work. What happened that day was one of the best examples of collaboration and innovation that Paul and I collectively achieved over the 25 years we worked together! While climbing about looking at the structure, one of us would come up with a different concept and the other would either expand on that idea or come up with a reason why that would not work. Then, we would either fine-tune the alternative method or find a solution to avoid the perceived problem. This interaction went on for hours and each progressive step was mutually resolved as we proceeded. In the end, we decided to retrofit the entire bridge from the bridge deck itself and not use any barges, tugs or waterborne equipment. Our motivation was certainly not solely altruistically related to the environmental impacts, but we believed that the waterborne method would be very expensive and our method would be significantly quicker and

37 The waterway is not really tidal because it is not affected by the moon, but the water level changes dramatically (6'–8') through the seasons, depending on flows from Harrison Lake, and more predominantly, the volume in the Fraser River, which causes water to back up into the Harrison River.

cheaper. All we needed to make our ideas work was a special truck crane to perform the work with reduced extension of the outriggers to match the existing girders on the bridge, somewhat like we did on a smaller scale on the Hyland River Bridge. After that brainstorming session on site, we went over to the Sasquatch Inn, which was very close by and had a great lunch and a couple of pints to fine-tune our plans. (I must have skipped church entirely that day!)

Our next move was to calculate the weights of the various lifts and radii required to determine what size machine we would need. I then spoke with several crane manufacturers to see how much trouble it might be to custom-build the outriggers for this purpose, and it turned out to be no problem at all. All they had to do was install an additional stiffener in the outrigger beams at the correct location to suit our overall required width of the outrigger pads. All we had to do was order the crane and they would build it exactly to our requirements.

When the tenders were opened on April 15, 2005, there were only two bidders, which was quite unusual. We were the low bid; however, we were concerned that the Ministry may not award the contract because there were only two bidders,[38] but five days later, we received the formal letter of award. One of the first things I did was finalize our proposed erection plans (the special use of the truck crane) with Somerset Engineering and start the process to get approval from the Ministry for the changes in erection procedures. The process of working out subcontracts for the supply of structural steel and steel deck grating was also an immediate priority after receiving the contract award. A preconstruction meeting was held on May 10, 2005. The Ministry accepted our proposed changes to the erection scheme, and we ordered the new custom truck crane for this work on July 26, 2005, from Falcon Equipment, who was located close to our office in Port Kells. Delivery of the crane was expected early in the fall, which would work perfectly for our planned work schedule at site.

One of the other contract stipulations was to employ flag persons 24 hours per day on each end of the bridge whenever one of the traffic

---

38 Normally, on a bridge project of this magnitude, we would expect to see between six and 10 bidders, but we thought that many of the regular bidders declined to bid due to the difficulty and complexity of this job.

lanes was closed, which amounted to almost the entire duration of the project because not a great deal of work could proceed without one lane closed. This flagging cost amounted to approximately 20,000 person-hours based on $35.00 per hour, and without considering overtime or vehicles, the cost would be over $700,000! Perhaps this stipulation was intended to be a make-work project for the region, but not thinking anything like a socialist, we thought there had to be a cheaper and more efficient way to do this, so we developed an alternative procedure for traffic control. Again, we worked closely with Tom Bayntun and the District Manager, Barry Eastman on this proposal and responded to each and every concern they voiced, eventually resolving all of their issues. The alternate traffic plan included the installation of the following items on each end of the bridge: a programmable message board; advance overhead warning flashers; street lights and concrete no-post barriers on the approach to the bridge and on the stationary spans; an overhead traffic signal system; a set of programmable traffic arms; and a visible timer display so motorists could see when the lights would change, which discouraged queue jumping. Special water-filled barriers were used at the edge of the traffic lane on the swing span of bridge and could be drained to move when necessary. Fraser City Installations supplied and installed all of these traffic control services. This extensive traffic control system worked beautifully and still saved hundreds of thousands of dollars.

Rapid-Span, from Vernon, B.C., was awarded a subcontract to supply all structural steel. Because we were matching new girders to existing steelwork, we had to accurately perform as-built measurements of the existing girders and all bracing bolt locations in order to provide that information to Rapid-Span so they could fabricate the new girders to accurately fit up in the field. I remember Diane Lemieux and Dennis Case were very involved in this painstaking process. Cambers of the new girders were fabricated exactly as noted on the shop drawings for the original girders made by Dominion Bridge, so once fully loaded, they should match the existing girders. Interlocking Deck Systems International (IDSI), also located in Pittsburgh, was awarded the supply contract for the steel deck grating system. Although I enjoyed visiting

Pittsburgh for the Hyland River Bridge, this was our third bridge with a similar type of deck grating, so we knew what to be looking for, and IDSI's subcontract was very specific to avoid the unfortunate issues we experienced on the Churn Creek Bridge. Benco Fabricators was awarded the subcontract to supply all of the miscellaneous iron, including the deck joints and the steel parapets on the swing span.

Preparatory work started on site about June 1, 2005, with the installation of traffic control equipment, a complete suspended work platform under the eastbound traffic lane and fixed decking under the centre two bays of the steelwork for the length of the stationary spans. After that was completed, we installed the barriers that ran down the middle of the bridge and restricted traffic to one lane in September 2005. Our demolition crew could then remove the steel curb, guard railing and expansion joints and sawcut the deck over the exterior girder into panels that we could lift out with the new 26-ton truck crane. The general method was to start at the swing span and work each way towards each abutment with the crane only working directly over its rear outriggers. Strengthening and painting the existing structural steel also started at that time. The eastbound exterior girders were removed in December 2005 and loaded onto highboy trucks on the traffic lane immediately adjacent to the truck crane. New exterior girders were then erected in a similar manner starting in late January 2006.

Harrison River Bridge - Installing New Deck Panels. Note New Exterior Girder on Left

Once the steelwork was completed, we were able to remove the remainder of the eastbound deck panels working from the centre span towards each abutment. I recall that a local farmer took all of the old deck panels from the bridge site and used them for riverbank bulkheads, as he did with the panels we removed on the Fraser Hope Bridge about 10 years earlier. Once the old deck was removed, we installed new Nelson studs to all of the existing top flanges and coated the top flanges with a corrosion-resistant product. We then installed the new steel deck grating, starting at the abutment and working off the new decking, progressing out towards the swing span. Key tradespeople in Art's crew on this project included Joe Mather, who became the general foreman after finishing up another job at the Coquitlam Dam Seismic Upgrade; Rick Johns, who ran the boom truck; Dennis Case, who supervised all of the structural steelwork, steel decking and reinforcing steel; and Wayne Clark, Jari Mackinen, Marc Farley, Glen Mosely, Perry Point and Korey Mather.

Harrison River Bridge - Installation of Deck Panels

Once all of the steel grating was installed, along with the new deck joints, we placed the deck overlay concrete. Because the steel deck grating was equipped with metal pans on the bottom, all we had to do was fill the grating with concrete and this did not require the use of a deck-finishing machine. After the deck concrete was completed, the next task was to construct reinforced concrete parapets for the full length of the stationary spans. Due to the time necessary to form, pour, strip and concrete finish the parapets conventionally in the field, we proposed another alternative process that we felt would look better and cut the time involved by at least 50%. Some months earlier, we proposed to M.O.T.H. to manufacture the parapets in a concrete precasting facility and install them in the field in ~10' lengths. The plan was to set them accurately to grade using shim stacks and injecting grout beneath the parapets after setting them in place. We proposed anchorage for the parapets with the use of anchor bolts that would be accurately cast into the base of the parapets and would extend through predetermined holes in the grid deck with a steel plate and double nuts below. In addition, there would be rebar dowels extending out of the grid deck concrete and into block outs in the parapets that would also be grouted in the process. This plan was loosely derived from a method used by Public Works Canada and Sandwell Engineering on some bridges on the Alaska Highway that I was aware of and was somewhat similar to the parapets we recently installed on the Hyland River Bridge. We retained Dragan Majkic from QR Engineering who provided the design calculations and stamped drawings for this alternate anchorage that met the very stringent specified PL-2 crash standard design requirements. Once this alternative method was accepted by the Ministry, we negotiated with IOTA Construction from Chilliwack to produce the concrete parapets using high-performance concrete, galvanized reinforcing steel and anchor bolts with the use of our steel parapet forms erected upside down in their shop. This method of casting provided a very smooth, durable finish to the concrete surface and enabled IOTA to perform a much better curing process in their facility than we ever could in the field.

Installation of Steel Parapets on Swing Span with Precast Section to the Left

Once we began installing the precast parapets in the field, we accidentally broke one of the 25 m (1" diameter) galvanized rebar dowels coming out of the deck with one of the outriggers on our boom truck, but the crew noticed that it broke far too easily. When they investigated other similar dowels, they found they could literally break these dowels off with their bare hands and that certainly should not be possible! I received a video at the office showing Kyle Jones literally breaking these dowels off by hand! Something was seriously wrong here and this problem commanded a great deal of my time for the next several months and involved much suspense and intrigue. The result of this strange incident included several scientific investigations and several companies involved were exposed to serious potential liability. Initially, most people blamed the galvanizing process as the cause of such brittleness in the steel. However, through exhaustive analysis of the galvanizing processes and testing by at least two consulting firms that we engaged, it was eventually determined that the rebar dowels were overstressed by cold working[39] that size and grade of rebar around too small of a pin in the bending process, causing strain age embrittlement. This overstressing was possibly made worse in the hot galvanizing process, but galvanizing alone did not cause this to happen. The eventual solution to this problem was to stress relieve the remaining rebar after bending, by heating it in an oven at 1200°F for a period of time (something like six to eight hours, as I recall) prior to galvanizing, and this was the end of the problem. Heat treatment causes the molecules in the steel to realign and therefore, regain significant strength. This incident ended up being a serious learning moment for us and everyone else involved, including the steel manufacturer, the rebar designer, the rebar supplier, the galvanizer, the precast company and the Ministry. This experience cost approximately $100k to resolve, including engineering fees, testing charges and job delays. Fortunately, all of the parties involved were very forthcoming and quite generous in accepting the extra costs, so the matter was concluded amicably. I clearly remember that Rick Froese at Fraser Valley Steel and Dominic at Lower Mainland Steel were particularly good to work with through this entire process.

With the eastbound lane completed successfully and well under our budget, we moved traffic over to the new deck in the middle of July 2006 and went through the entire process again on the westbound lane, without any major incidents. That lane was completed by November 2006 and both lanes were reopened for the public. The polyurethane wearing surface was eventually completed by SRM Inc.[40] in the summer of 2007.

This was another one of those very pleasant jobs to be involved with because there was excellent cooperation from everyone when challenges or problems arose and they were all resolved quickly and amicably. The Ministry was particularly good to work with and both Art and I got along with Tom very well. Overall, I believe the Ministry was very happy with our work, including the minimized inconvenience to the public. Buckland & Taylor were also fairly receptive to our alternative proposals, and we were

39 Cold working is the process of bending steel without the use of heat, which is common for bending rebar.

40 SRM was owned by Philip Sutton-Atkins, who I attended North Delta High School with.

able to stickhandle around any of their particular requirements without too much trouble. After all, there is a solution to every problem. Paul and I often looked back on that job with pride due to the significant changes in methodology we were able to orchestrate that ultimately saved a great deal of money and time to complete the work.

WVCC 20th Anniversary Party

WVCC held its 20th Anniversary party in the Grand Hotel, Kelowna on the weekend of September 21, 22 and 23, 2006. All staff and their spouses were invited for the weekend, and we all had deluxe rooms, fine dining and free-flowing beverages all weekend. The planned events for that weekend included a go-cart racing competition, a wine tour, a mountain bike ride down the KVR rail grade down from Naramata to Penticton, a luncheon at the Merganser Restaurant and a "tacky tourist party" on an evening boat cruise out of the Kelowna Yacht Club. I know that Kyle Jones won the trophy for the fastest elapsed time at the go-cart track races, and I got the highest average speed including all time trials—one of the prouder achievements in my life! I also remember that most of the guests conspired against me at the tacky tourist party by wearing socks under their sandals because I might have done that once or twice on a hot summer day in the office, but I couldn't quite bring myself to wear bare feet and sandals to work. Obviously, the entire event was a lot of fun and everyone thoroughly enjoyed the entire weekend. It was a great way to get everyone together from all over the province and celebrate 20 years of hard work, success and friendship.

Arnold Palmer

Around this time, Gladys and I travelled to Palm Springs on an annual basis for at least a few weeks, renting a condominium, playing a lot of golf and tennis and going out for great dinners (when we didn't cook for ourselves). Many of our friends also travelled down every year, so we had lots of fun with the Mannings, Burns, Copleys, Atkinsons, Vanlerbergs, Talbots, etc. On one occasion, we played a round of golf with Drew and

Janet Copley at the Golf Club at Terra Lago in Indio, and we all enjoyed ourselves very much. After the round, and after I paid Drew the nominal $6 for the standard bet, we all went out for lunch at the Beer Hunter in La Quinta, which is a big sports bar with about 40 big-screen televisions. (This is a great place to watch the Superbowl or the Phoenix Open, which I have done a few times.) Anyways, we were sitting there waiting for our lunch to arrive when I noticed someone walk in with a small group and sit down just a couple of tables away. I nudged Gladys and Drew and said, "Don't turn now but Arnold Palmer just walked in and sat down over to our left!" Well, this was something. I sat there for a moment and then remembered another opportunity I had to meet a famous golfer. Several years earlier, Gary Player walked by me in the Marriott Desert Springs Hotel lobby in Palm Springs. I still regretted not just taking two steps and saying hello to Gary, so I was not going to let this opportunity to meet Arnold pass. Arnold's table had not yet ordered drinks, so I got up, went over and said hello to him. Arnold was as friendly and gracious on that day as we always heard he was. He put his hand out to shake mine and he clamped his other hand on top of mine in one of the warmest handshakes I have ever felt! We spoke for a couple of minutes, and he was so friendly and accommodating that I didn't want to take advantage of his good nature. I was certainly impressed. Gladys and Drew went over a few minutes later and Arnold autographed our golf score card, which was an Arnold Palmer course that we played that day. One thing I noticed about Arnold, and what Gladys later remarked about him, was how good he smelled with his cologne. Having never worn cologne other than on very special occasions, I have been wearing cologne every day since that day!

Aberfeldie Redevelopment Project - Cranbrook, BC

The Aberfeldie Redevelopment Project was a contract with BC Hydro in the final amount of $38,154,400 that was tendered by WVCC on October 11, 2006, for the complete redevelopment of the existing hydroelectric facility on the Bull River, east of Cranbrook, B.C. The existing facility was a 5 MW plant that was constructed as a run of river operation in 1922 with a 27-m-high concrete dam later constructed in 1953 to provide water storage. In addition to the concrete dam, the existing plant included 1 km of ~ 3.65 m diameter wood stave low pressure penstock (LPP), a large wood stave surge tower about 15 m in height, a riveted steel high-pressure penstock (HPP) and an existing powerhouse equipped with twin 2.5 MW turbine generator units. The total head of the project was 84 m. Interestingly enough, the site of the dam is adjacent to a canyon where in the early 1900's, an old log flume carried logs down to a mill that was located on the Bull River, not far from the existing powerhouse. We found several historical remnants of this operation on site, including what looked like an old, narrow gauge steam engine buried in the weeds and overgrowth.

The contract scope of work for this project included the complete demolition of all components of the existing facility except the concrete dam, where we had to make major modifications in order to accommodate a new twin intake structure, as well as the reconstruction of the LPP, the surge tower, the HPP and a completely new powerhouse that housed three 8.3 MW turbine generator units and an emergency bypass valve. It was actually a beautiful project at a very beautiful location not far from the base of the Rocky Mountains to the east. It was right down our alley work-wise, and we had a lot of experience with this type of project.

Doug Baker was the project manager for BC Hydro, working out of the Burnaby office, and Jim Horkoff was the site supervisor. Around the time of tender, BC Hydro was still working frantically to get final approvals for the project in order to proceed with the BC Hydro Board of Directors and the BC Utilities Commission, so we agreed to extend the tender period from November 24, 2006 to February 12, 2007, when the award was finally issued.[41] Doina Dobre, who had recently completed a good-sized project for us with the Greater Vancouver Water District for the seismic upgrade of the Coquitlam Dam, was initially the project manager on site, along with John Carvell and Kyle Jones as area superintendents. Diane Lemieux and John Turecki were both project engineers on site. A short while later, once he was finished at the Harrison River Bridge, Art came in to replace

41 With the benefit of hindsight, we should have run when BC Hydro could not award on time. It would have saved us a lot of money, stress and countless sleepless nights!

Doina, who was relocated to head office to assist with job administration from there. Bob Lamond was hired in April of 2007 as a dedicated project safety manager. Gary Margesson and Russ Dowdeswell started on July 3, 2007, as additional area superintendents as our work areas expanded. The surveyors on this project included Bruce Balfe, James Lattimer and Dean Shipytka, who also managed a lot of the quality control.

Once we finally got the project award and mobilized to site, I invited Jim Horkoff out for dinner one night in order to introduce him to Art. Jim would not let us buy him a $20.00 dinner because he thought it was morally unethical! We should have seen the writing on the wall that night. The project got off to a rough start as the required submissions for BC Hydro were exhaustive to say the least and it appeared BC Hydro was actually trying to obstruct any progress at all with the design drawings lagging behind. In the past, we had been deeply concerned about working with BC Hydro because their reputation for being very difficult was well known throughout the industry. However, they told us in the bidding stage that they wanted to build this project as a model for future historical asset upgrades, much like smaller private hydro developers managed their projects (IPPs). That is exactly why they outsourced the complete engineering for the project to Knight Piesold, who was very experienced in private hydro developments, even though BC Hydro already had hundreds, if not thousands of engineers on staff already. Well, it could not have been managed any differently than a private development! We had already built six IPPs in the province by that time and this was nothing like anything we had seen in the past. For an illustration, the Soo River project cost the owners approximately $1.5 million per MW output to build the entire greenfield project, whereas BC Hydro spent $95 million to produce 25 MW to upgrade an existing facility, which was $3.8 million per MW!

One of the first things we had to do once we got to the site was construct an earth fill berm entirely around the locations of both the existing powerhouse and new powerhouse, and this berm was partially located in the wetted perimeter of the Bull River. The first challenge was to locate a suitable clay material for the core of the berm, to the satisfaction of BC Hydro, which we eventually did but not without a great deal of scientific analysis, hand-wringing and several expert opinions from various levels of BC Hydro management. Extensive testing of various rip rap materials throughout the East Kootenays was also undertaken to satisfy BC Hydro and that source was finally located after a great deal of geological and testing analysis. In order to reduce any possible environmental impacts to the river during this operation, we installed an Aquadam[42] along the edge of the river to isolate our entire work area. This operation did not work that well due to the flows in the river, even though it was the middle of winter. Eventually, we were able to get the berm installed, but this was only the beginning of our problems with the excavation inside of the berm.

Currier Contracting, a local Cranbrook company, was given a subcontract to perform all wood stave penstock and surge tower demolition, along with the complete demolition of the existing powerhouse. Eventually, once their work was advanced far enough ahead of us, we got started on the required demolition at the intake and rock excavation along the low-pressure penstock. Little Rock Drilling, a local contractor, performed the drilling and blasting as a subcontractor for us. The new low-pressure penstock incorporated 850 LM of 3.35 m diameter Weholite HDPE pipe and 150 LM of 2.90 m welded steel pipe, which was supported by concrete foundations over gullies along the upper penstock alignment. The Weholite HDPE was manufactured by KWH Pipe out of Saskatchewan and was fusion welded in place with a special machine we rented from them, along with welder training that they provided our crew. The steel pipe for the LPP and the HPP was manufactured by T Bailey, located in Anacortes, Washington. After working with several pipe manufacturers over the last number of years, I found T Bailey excellent to work with, and they were a highly professional group. Steve Siu, our welding engineer, again worked closely with us and T Bailey, as well as Vic Bilodeau for the field welding. Gene Tanaka was the president of T Bailey and, along with Tim Sextant, Randy Scott, Dave Edwards and Mike Whittig, he was on top of all the necessary details. We also worked very well together on this project; however, the continuous steel, welding and coating inspections by various parties at BC Hydro

42 An Aquadam is a proprietary product that is basically a long sausage of webbed plastic—this one was about 12' wide and perhaps 100' long. It was located on the river's edge and was pumped full of water to create an isolation dam between the river and the edge of the berm that we would build.

were a complete pain in the ass! T Bailey's scope of supply and fabrication included a very large diameter steel wye at the intake that bifurcated the flow from two separate intake structures into the single LPP and a large manifold of elbows at the powerhouse to feed the three turbines and the emergency bypass valve.

The high-pressure penstock involved 150 LM of 2.74 m diameter welded steel pipe that was surface mounted on concrete pedestals down the steep incline to the powerhouse at a grade of 84%. This was far steeper than any other penstock section we had installed in the past. (Miller Creek was as steep as 73%). The initial task on this steep slope was to perform the necessary excavation and machine scaling of the rock from top to bottom and this operation was extremely difficult due to the steepness of this shear rock slope. To accomplish this work, we had an excavator start at the top and work its way down, slowly excavating and gathering any loose material off the slope while sitting on top of its own muck pile. I think this was Gord Trottier again and his son, Brian, who worked their way down with a 300-sized excavator that was at times connected to a winch up above. As the excavation moved down the mountain, we could then drill and blast out the foundations for the HPP pedestals by winching an air track down from the top. As this steep slope excavation and machine scaling was completed at any given height, we then had drillers install rock bolts and welded wire mesh with the aid of a helicopter to stabilize the slope. Eventually these operations were completed down to the base of the slope and our carpenter crew could build stair access to all of the HPP pedestals and the formwork and concrete placement could get underway.

Aberfeldie Hydro - Preparing High Pressure Penstock Foundations

In the meantime, we commenced the powerhouse excavation, which was located along the edge of the Bull River inside of our protection berm. This excavation proved to be extremely challenging due to the artesian water pressures in the silty, sandy native material that we experienced throughout this operation as well as the fact that we were working about 20' below water level. Notwithstanding the extensive lengths we went to locate and place the optimal material to satisfy BC Hydro's experts for the core of the berm, we experienced boiling sand from the bottom of the excavation and piping that appeared to be coming from the river below the berm, which are both potentially dangerous conditions. To overcome these conditions, we had to employ virtually every possible dewatering method to eventually achieve a reasonably dry hole where we could start the concrete foundation work of the powerhouse. These dewatering methods progressively included the use of deep wells, shallow wells, numerous piezometers, relief wells, a complete well point system, along with surface ditching with sumps to pump from—you name it, we had it! To deal with the underground piping from the river, we installed a complete steel sheet pile cut-off wall down the centreline of the protection berm and also ground filters using a number of differing grain-sized granular materials combined with geotextiles to stabilize these areas and prevent further piping and erosion on the inside of the berm and the face of the excavation. Glenn Skaar from Canadian Dewatering assisted us as much as possible with all of the dewatering methods and equipment, but it became a hugely expensive undertaking. We initially carried a budget of $150,000 for the dewatering, but we spent many times that number when all was said and done. In the end, we even hired John Balfour, a hydrogeology specialist I located from Vancouver, to examine the site and assist us in determining the best scientific dewatering methods. What he believed was happening was that the underground flows were actually coming from the uphill side and then through the underlying soils, rather than directly from the adjacent river, as one would intuitively think, and hence, where we had concentrated most of our efforts. We eventually got things under control enough to get the foundation pours in, but it caused endless sleepless nights and a whole pile of money. The remainder of the powerhouse substructure was built over the next number of months while we continued building the intake structures and installing the HPP.

The installation of the HPP on the steep rock face was certainly one of the most challenging tasks that we faced on this project, but we worked closely with Somerset Engineering again and devised a very unique methodology for the installation. KWH Constructors performed the pipe operation for the HPP for us as a subcontractor on an incentive-based, cost-plus-to-a-target-price contract model. Basically, we installed two pairs of large, custom-made rollers at each concrete pedestal on the hill and then winched the pipe upward every time we welded on a new pipe section at the bottom. We had a 230-ton Manitowoc crane located at the bottom with special rigging adapted to hoist the pipe sections on the correct slope as we added them. Each time a pipe was added at the bottom, the circumferential weld was completed, weld testing performed and the joint coating repaired both inside and outside the pipe. The welding and coating were performed inside a weatherproof structure that we built in place so this work could proceed in any weather conditions. Vic Bilodeau, who did all the welding at Miller Creek for us, was again our welding subcontractor, and he and his crew were excellent to work with and performed the welding quickly and efficiently.

A large stationary winch was set up at the top of the hill to pull the pipe up the hill on a multi-part line and a long, lattice steel type nose was fabricated and installed onto the uphill end of the pipe to reduce the pipe's deflection over the long spans between pedestals. One of the problems with this general method of installation was that because the pipe was fully coated, the rollers had to be specially designed to carry the full weight of the pipe but also had to be fabricated from soft enough rubber to avoid damaging the exterior pipe coating. (The access was so difficult on this steep slope that we could not afford to do any coating repairs on the hill.) The other problem we had to overcome was the circumferential ring stiffeners located along the length of the pipe, which prevented the pipe from moving over the rollers. To accommodate this issue, we installed the rollers in pairs so when the ring stiffener approached the first set of rollers, we would remove them while the pipe remained supported on the second set of rollers. We would then advance the pipe up the hill a few feet more, allowing us to re-install the first set of rollers and remove the second set to allow the stiffener to pass by. This procedure had to be repeated at each pedestal support for each ring stiffener, which added quite a bit of additional work to the process, but the crew soon became fairly efficient with these operations, utilizing the pinned connections on the roller supports. Even though this was an extremely difficult penstock to install due to the above challenges, this method proved to work quite well.

Aberfeldie Hydro - High Pressure Penstock Installation at 84%

Overall, the powerhouse involved a total of 4636 m^3 of concrete and much of that work was delayed into the winter months due to a number of factors, many of them out of our control. Winter heating and hoarding was necessary for most of the concrete. Our powerhouse foreman was Joe Mather and in his crew were a number of long-term employees, including Wayne Clark, Glen Mosely, Jari Mackinen, Mark Farley, Korey Mather, Ryan Young, Wayne Zorn and a few local workers. Gary Gallagher set up and ran our Fastway concrete plant on site and we produced a total of 9275 m^3 of concrete for this project. Prince George Steel supplied all of the reinforcing steel and AJC Contracting (Darryl Bohmer and company) placed all of the reinforcing steel on this project as a subcontractor. I am pretty sure these were the same guys who placed my rebar at the McPhee Bridge many years earlier. Westpark Electric again performed all of the electrical work on site including the installation of all turbine generator electrical as well as the complete balance of plant in the powerhouse. VA Tech supplied and installed all of the turbine generator equipment directly for BC Hydro. Nothing was easy on this job but the electrical scope went relatively smoothly considering how tough it was for the general contractor on this site. Fabrite Services, a talented local steel fabricator supplied much of the miscellaneous steelwork for the project including walkways, hand railings, steel grating and the steel surge tank, which was fabricated and erected by their crew in circumferential rings. Mike Kozinuk was the president of Fabrite and our day-to-day contact was his son, Cole Kozinuk. They were a great company to work with—they certainly understood the difficulties we experienced with BC Hydro and worked hard to meet our needs.

The biggest problem we had with the powerhouse substructure was that the quantities of work kept changing dramatically. Also, the complexity of the work and the time required to build the work increased as the quantities of work increased, but BC Hydro would not provide an extension of time, so this amounted to a full-blown acceleration of this work. Two prime examples of these increases were the concrete quantity, which went from 3449 m^3 to 4636 m^3 (+34%), and the reinforcing steel, which increased from 379,400 kg to 791,076 kg, an over double increase of 108%. So, based on the tight confines of this powerhouse, when your rebar quantity over doubles, the man-hours required to install this also doubles. In fact, the labour required for the rebar installation went well over double because of the inherent increase in the complexity of the work and the tighter confines in which to place all of that rebar. For example, the rebar factor for the powerhouse at the time of tender was 110 kg/m^3 of concrete and it went to 199 kg/m^3! There was so much rebar in the forms, it was nearly impossible to place the concrete! Now, admittedly, when quantities of work increase in a unit-price contract, the contractor receives more money to deal with those quantities, but in this case, because the powerhouse was the critical path of the project, we required more time to complete the work and that extra time was refused by BC Hydro. A formal notice of delay claim was submitted on January 15, 2008, as soon as we became aware that our completion dates were in jeopardy.

One other example of a related argument we had with Jim Horkoff involved how he calculated the payment volume of concrete and deducted the volume of rebar from that total, which was the first time we had ever heard of that being done, and certainly this practice was not in accordance with CSA, the BC Hydro Specifications or any other accepted standards in the construction industry. A subsequent conversation ensued with a senior BC Hydro manager, Tony Cohen, during which he stated he had never heard of deducting rebar from concrete payment quantities in 41 years in the industry! This was worth about $63,000 to us, and we eventually were paid this amount but not without a great deal of argument and unnecessary stress. There were dozens of seriously problematic items like this. They arose almost every day on this job and it was abundantly clear that for whatever reason, Jim Horkoff tried everything possible to make things difficult for us.

In the end, once the work was completed in the fall of 2008, we assembled a first-class claim with the assistance of both Mike Demers and Al Morgan from Revay and Associates. This powerhouse delay claim requested an extension of time of 143 days along with additional compensation for the constructive acceleration, winter work, increased complexities and extended overheads. BC Hydro also assessed liquidated damages of $10,000 per day against WVCC for the alleged delays in completing the work, which were also strongly disputed. After a year and one half of correspondence, including upwards of 100 documents, numerous meetings, discussions, legal wrangling and negotiations, BC Hydro finally agreed to pay us $3.3 million for full compensation to cover the various delays and the resulting acceleration and to eliminate all liquidated damages it had assessed. The process was a long and painful one that included appealing through several steps in the hierarchy of BC Hydro management right up to the CEO, Chris O'Riley; nonetheless, a settlement agreement was finally executed on September 2, 2009, some 20 months after submitting the initial request for an extension of time. This was, unfortunately, one of those jobs where we received very little collaboration or cooperation from the owner, and in particular, from Jim Horkoff, so it was no fun for anyone, including the hourly tradespeople. Jim, who had come up the BC Hydro ranks from a teamster, then safety officer, loved reciting the general conditions of the contract in every piece of correspondence he penned, and these were known as FAM's, which is BC Hydro's acronym for Field Advice Memos. Jim wrote hundreds of FAM's on this job, and he would recite a minimum of one general condition per FAM, no matter what the topic was intended to be. After spending a life in construction and construction management, it is my humble opinion that the contract general conditions incorporated into construction contracts are usually written by lawyers in advance of any project in an attempt to try to deal with any number of situations that might arise during the execution of the project. However, the general conditions are not an effective way to manage a project on a day-to-day basis or live your life. But this was definitely Jim's life! We had been running a successful construction company for 23 years up to this point, but this was, unfortunately, a low point in the company's history. At the end of this project, we had completed a very difficult job and we were all proud of the end product, regardless of the problems we encountered.

Aberfeldie Hydro - High Pressure Penstock Down to Completed Powerhouse

The Sale of WVCC

Once the Aberfeldie Project was completed and the claim eventually accepted and paid by BC Hydro, we were all collectively tired and somewhat deflated. The resolution of our claim took a lot of effort and time from both Paul and I, as well as the rest of the staff, and the biggest contract we undertook following Aberfeldie was a highway bridge seismic retrofit that

Art did for M.O.T.H., valued at $1.4 million, as well as a couple of smaller bridges. Well before the final resolution of our claim with BC Hydro, Paul and I began to think seriously of selling the company as an operating entity, and we knew that was likely going to be a long, taxing process after being in business for almost 25 years. Paul eventually hired a business consultant to market the company for us, and they brought in lots of tire kickers but no one who would pull the trigger. We entertained many suitors over the next year, including some of the largest construction firms in Western Canada. However, one guy who seemed very interested in buying WVCC was Bryan Hall at Tyam Construction. I had known Bryan for years, and I met with him on a few occasions to discuss this possibility and our conversations progressed nicely. Tyam seemed like a decent match to me as they had recently completed a large volume of heavy construction work on the new RAV Line project, which was a new rapid transit project that ran down Cambie Street to downtown Vancouver. It was Bryan's vision that WVCC would expand Tyam's diversity in the construction industry and bring a lot more heavy-civil structures work such as railways, mines, bridges and hydroelectric projects onto their books. Tyam had recently taken on two financial partners to assist them through the RAV Line project, which gave them the additional financial horsepower to take on more, larger projects, and I thought that would suit our continued operations as well. In my mind, if WVCC had a single operational fault, it was a lack of financial and administrative capacity to take on more than one larger-type project (which we had become experienced with) at any one time. This capability would take out the highs and lows of our cyclical cashflows as well as potentially create better utilization of our administration, senior staff, and owned equipment. I met with Bryan on a few occasions at Earl's Restaurant and Brown's Social House in Langley to discuss the proposed transaction, and I was enthused to see the eventual culmination of this sale agreement. We eventually agreed on an equitable price and were in the middle of finalizing this framework sale agreement when, all of a sudden, on about March 10, 2010, Bryan was terminated from his position at Tyam by the three other partners for his alleged personal misuse of company funds. This certainly put the kybosh on our pending sale. Or did it?

Interestingly enough, Bryan never sat around very long after this shocking event of being fired from a company that he started, and he immediately formed a new company called Hall Construction along with his father, Dennis Hall. Bryan again approached me and said that he was still committed to purchasing WVCC and Hall Construction would honour the details of our previous sale agreement. Shortly after we renewed our negotiations, Bryan received a letter around mid-April 2010 from the directors of Tyam Construction, his former partners, stating he could not proceed with the purchase of WVCC as he would be in breach of his fiduciary obligations to Tyam because he became aware of the opportunity to purchase WVCC while he was a director of Tyam. This legal wrinkle was of serious concern to Paul and I as this was clearly a direct interference in our sale. After a few days, I wrote a letter to the Tyam directors and essentially told them we already had a sale pending to Hall Construction and they should either "shit or get off the pot!" (I may have used some more formal business language at that time, but that is exactly what I meant.) Several days went by, then I received a letter from Tyam responding that they would like to proceed with the purchase with similar terms and conditions as the earlier proposed sale to Hall. Well, that was interesting. Bryan also confirmed to me that his hands were tied and if Tyam wished to buy WVCC, then he had no ability to do so without the risk of being sued and dragged through a protracted legal mess with his former partners, which he clearly wanted to avoid.

The directors of Tyam were Jason York, the original partner of Bryan's when that company first started, along with Rob Barker and Ralph Jordan, the two businessmen who earlier provided financial support to Tyam and essentially took over the majority of control of Tyam with Bryan's departure. They had hired Mike Luers as an executive vice president of Tyam to take over the day-to-day management of that operation, and Mike became my direct contact in the negotiation for the sale of WVCC. Although Mike had a background in chemical engineering in the field of pulp and paper, I thought he adapted quite well into the executive level of management of the construction business and was keen to learn everything he could about this industry. Mike and I got along very well, met many times to work out the various details of the purchase and were working positively together to complete the sale. One aspect of the sale that was important to Paul and I was to sell the shares of WVCC, not just the assets, in order to utilize the capital gains exemption of $750,000 that was available to each of us for a share sale. This meant that the first $1.5 million of the sale would be paid to us personally without incurring any taxes, which was a huge benefit. However, if Tyam was to purchase the shares of WVCC, it would be effectively purchasing 50% of Westpark Electric, as WVCC owned 50% of Westpark. Tyam was resistant in purchasing 50% of Westpark because it was not a core business interest to them, and they stated they knew nothing about electrical companies. This was again a very concerning turn of events for Paul and me. This was also quite surprising as Westpark had established an excellent reputation in the industry, and they were always very profitable, even more so than WVCC had been for the last couple of years. Upon further reflection, and no doubt several discussions with lawyers and accountants, I met with Mike Luers again early one morning at the White Spot in Walnut Grove and offered him a proposal that included an "earnout agreement" for the sale of the Westpark shares where the price they effectively paid for those was solely based on the company's profitability over a period of five years. This agreement set a range of five annual payments to Paul and me in amounts between $0.00 and $500,000, depending on the profit picture of Westpark each year. This scenario looked great to Mike as Tyam would only pay for proven profitability in which they would technically participate at a rate of 50%, so they could 'theoretically' pay for half of the company on their share of its profits. He readily sold this agreement in principle to the three remaining shareholders of Tyam. This proposed sale went through as smoothly as it could have possibly gone, but the paperwork involved was unbelievable! Lawyers and accountants from both sides were involved for quite some time, and Robin MacFarlane was rigorous in managing the legal affairs on behalf of Paul and I. Robin was always the consummate professional in these matters. Lesley Huygen, our WVCC accountant, was also very instrumental in the sale, constantly updating our financial reports and pro forma calculations to analyze every nuance of the sale all the way through the process. My own personal accountant, Paul Martin (not the federal minister of finance, but maybe just as smart!) was also critically involved and was extremely helpful in devising the best methods to minimize the tax effects of this sale for both Paul and me. A subtle, but key component of the sale agreement we stipulated was that Tyam's shareholders would have no voting rights with the Westpark shares until they were fully paid for, which was acceptable to Tyam's shareholders. (More on this later.)

One other stipulation of the sale agreement was that all employment obligations had to be concluded by WVCC prior to any employee starting work for the company under the new ownership. That requirement was necessary to avoid transferring any liabilities for severance pay for long-term employees to the new owners. That in itself was quite a costly exercise because we had so many long-term employees, and in my own case, I was paid a retirement allowance in the amount of $180,000, having been an employee for 24 years. Also, because I effectively sold the company, we did not have to pay any commission to the business consultant, and Paul and I agreed that an equitable commission should be paid to me instead, which was considerable. The final sale was finally inked on July 31, 2010.

The sale was predicated on me staying on to run WVCC for a period of two years, but Paul was able to negotiate his immediate retirement. As Paul was a 70% shareholder of WVCC, he received quite a nice sum of money through the sale, and along with the fact that he owned the building we operated in for the last 20 years, he was financially set and ready to retire. He appeared to me to be totally relieved to have the company sold and was really looking forward to retirement, although he was still involved in assisting at a high level with some estimating for WVCC over the immediate short term. As difficult and stressful as the last couple of years were for me, I was now somehow re-energized and ready to move the company forward through its next exciting chapter. I was only 56 years old and was far from done yet. We moved into Tyam's huge building in Aldergrove, and I can tell you it was quite a major undertaking to move all of our office furnishings, computer systems, warehouse inventories, company records, construction materials and equipment out to the new facility. All of the other long-term employees came along with the sale and we were collectively geared up for taking on some challenging new projects. Over the several months the legal and accounting details of the sale were being officially finalized, I engaged quite a bit with Mike Luers, Jason York, Rob

Barker, Scott Ashton (Tyam's accountant) and Lesley, developing a comprehensive business strategy for how we planned to operate WVCC over the next few years. We were definitely all on the same page when it came to wanting to pursue larger-type industrial projects and the concept of operating within joint venture partnerships to provide new, larger opportunities and additional resources while spreading out the risk. As a group, we drafted a comprehensive business and financial plan with pro forma charts that laid out our agreed operational plans for the company.

In the fall of 2010, WVCC was finishing up a reinforced concrete haul road bridge that we were building for Lafarge up in their pit operation in Egmont. Jim Mutter, an ex-Goodbrand employee and WVCC's former equipment manager, and Rick Wagner, also a former Goodbrand employee, now worked for Lafarge with whom we negotiated this contract. We also took on a number of Tyam projects that they were low bidder on, including some sound abatement walls along the Sea to Sky Highway in Lions Bay, a bridge deck in Langley and some concrete retaining walls at various residential and commercial developments they were involved in. Oddly enough, not that long after the purchase of WVCC was finalized and we became fully operational, the ownership group decided to wind up Tyam Construction and a lot of its equipment was sold off and staff dispersed. Should this have been seen as an omen??

Long Lake Hydroelectric Project - Stewart, BC

Around this same time, Regional Power put out the Long Lake Hydro project for tender near Stewart, B.C. This was a great, private, 32 MW hydroelectric facility that we had provided Regional Power costing and feasibility studies for on at least two occasions prior, and considering our past experience with IPPs and working in Stewart, we were very well-suited for this project. At some point that fall, we were approached by EBC-Neilson, a joint venture of two companies from Montreal and Quebec City, who were very experienced in large hydro projects in Eastern Canada and wished to join forces with WVCC to bid some hydroelectric work in B.C. There was a shortage of this type of work in Eastern Canada, but there were several private hydroelectric projects out for tender in B.C. at that time, including the Dasque and Middle River Hydro Project near Terrace, Kwoiek Creek Hydro near Lytton, Forrest Kerr on the Iskut River in Northwest B.C. and the Long Lake Project. EBC-Neilson was interested in bidding all of them. This arrangement suited our business plan perfectly, and Mike Luers and Rob Barker were fully supportive of these collective opportunities.

Francois Groleau was the general manager of EBC and Denis Lepinay was the general manager of Neilson. We also became very involved with Donald Proulx and Jose Rochette from EBC and Dominique Hotte from Neilson. We all visited the Long Lake site in late October and everyone was keen to proceed with this very challenging project. We quickly formed a legal joint venture in B.C. called WEN, an acronym for WVCC, EBC and Neilson. It was agreed that WVCC would be the project sponsor due to our experience with this type of work in Northern B.C., our union affiliation with the Independent Employees Association, our relationship with subcontractors and our ability to locate and hire experienced tradespeople in B.C. We also had previous experience working with Regional Power at Hluey Lakes Hydro in 1993 and more specifically, working with David Carter, who was the president of Regional.

We agreed we would jointly bid the Long Lake Project and then have a complete bid review in EBC's office in Montreal from January 3–5, 2011. We had recently hired Rainer Kraft, who had come to B.C. with Bilfinger Berger to build the Golden Ears Bridge, as our senior estimator and Alex Morrison, who had spent years working with Westpro Construction (a similar company in B.C. to WVCC), as a senior project manager. At the time, Dan Cave ("Study Dan") was living in Whistler working as a construction superintendent, and when we approached him, he seemed keen to re-join us for some of this new hydroelectric work after recently completing another IPP in the interior of B.C. The estimating process progressed very well in our Langley office where Paul worked with Rainer and Alex, along with Jose and Donald, who assisted on a part-time basis. One of the first steps was for each of the three companies to independently perform quantity take-offs so we could compare quantities and then agree on the final quantities to include in our estimates, which went fairly smoothly. Next, each company produced separate estimates for the various components of the work so we could make direct comparisons between the estimates. During that initial bid review in Langley to compare those estimates, we found that all of EBC/Neilson's labour productivities were much lower than ours, though our productivities were mostly based on our own historical records for all of the hydro projects we had built to that date. The man-hour factors they used were sometimes two or three times what we carried, meaning that their labour cost estimates were two or three times higher than ours! This was quite concerning to us, but EBC/Neilson was generally happy to accept our estimated productions and the bid review continued on in this manner. Once these comparisons were completed, a few of us attended to the final bid review in Montreal as planned. We all seemed to understand the project scope and requirements very well and EBC/Neilson was keen to take on this assignment. One great thing about working with the members of EBC/Neilson was that they loved to go out for great dinners with fantastic red wines like burgundy and bordeaux. I have no idea how much these dinners cost or what the wines were worth, but they were sure a nice treat for me! (EBC/Neilson combined did more than $500 million in volume per year, so we had no problem letting them entertain us.)

After some time, following our proposal submission for the Long Lake Project at the end of March 2011, we began final discussions with Regional Power and travelled to Toronto in April to negotiate the final terms and conditions of the contract as well as the final financial security arrangements. Mike Demers attended those meetings with us and it took several long days to finalize this painful process with Regional Power's managers and lawyers present. Regional Power was owned by Manulife Insurance, a financial management/insurance company that didn't like to take on any risk. Accordingly, their initial contract terms and security requirements were purposely very onerous to a contractor, and we had to be very wary of the potential risks that we were, or were not, willing to accept. Mike was instrumental in levelling that risk in our contract negotiations as best as we could, and we were all eventually satisfied with the outcome of those face-to-face negotiations. Acting on behalf of Regional Power during these negotiations were several managers, including Jean-Francois Martel; Nick Dhillon; Chris Lambeck, the CFO; and David Carter, the president. After the negotiations were complete and the contracts eventually executed, Nick Dhillon approached me and said something to the effect that it was a pleasure negotiating with me, and he thought that I was very reasonable and not nearly as tough as my reputation. This was a shock to me because I had exactly the same compliment for him—it was his reputation for toughness that had been a big concern for me!

Speaking of onerous contracts, earlier that fall, I began some serious discussions with some other major contractors in B.C. that had recently worked for BC Hydro and found that my own strong discontent with the way BC Hydro operated, particularly at Aberfeldie, was widely shared by every single contractor I spoke to. This was a list of the who's who in the B.C. construction industry and we agreed there had to be something we could do to improve this terrible situation. One of my first calls was to Philip Hochstein, the executive president of the ICBA. He had heard the same comments from his member contractors, simply that BC Hydro was not a preferred owner for contractors to work for. In the meantime, BC Hydro announced it was embarking on its five-year plan to modernize its heritage assets to the tune of something like $500 million per year for the next five years. As a taxpayer and a ratepayer, I had to wonder what a huge premium BC Hydro was paying for its construction services because they were so unnecessarily difficult to work with. We then formed an ad hoc committee of contractors to come up with a plan to force change, and we were quickly joined by Jack Davidson of the BC Roadbuilder's Association and Manley McLachlan of the BC Construction Association and we got a lot of assistance from Dan Doyle, who was the Chairman of the Board of Directors of BC Hydro. Over the next year or so, this high-profile group met several times with the highest-ranking officials of BC Hydro right up to Chris O'Riley, the CEO. Our group agreed to retain our construction lawyer, Mike Demers, to represent us in these discussions with BC Hydro. Eventually, with the aid of government pressure and a wave of disgruntled industry leaders, BC Hydro agreed to revamp their construction contracts in order to seek more competitive pricing in the marketplace. This process took some time and carried on throughout 2011, but a new contract was eventually finalized with this committee's involvement. Although I never had any further experience working with BC Hydro, I understand that the contract language changes were an improvement; however, the problems at BC Hydro were also cultural and it looked like it would take some time to change that.

Over the winter of 2010–2011, WEN bid the Dasque & Middle River Hydro project as well as Forrest Kerr Hydro, which was a 195 MW plant for AltaGas that included a huge underground powerhouse requiring thousands of cubic meters of rock excavation, which was a specialty of EBC. Fortunately, after months of work and negotiations, we eventually declined on the opportunity to be awarded the Dasque & Middle River Hydro project because we would have to take on far too much risk with regards to questionable subsurface soil conditions. We made specific qualifications in our proposal to protect ourselves and share certain risks with the owner; however, the owner wouldn't budge and demanded we accept those risks entirely. That project was awarded to another firm and our worst concerns regarding the soil conditions along the penstock excavation came to be. This project became a disaster for everyone involved, eventually ending up in the courts. We were certainly pleased that we foresaw these untenable risks and refused to accept them. Unfortunately, we were not successful with our proposal for the Forrest Kerr Hydro project either, even though we, and most people in the business, were convinced that WEN was the preferred contractor, mainly due to EBC's extensive experience and specialized equipment and labour availability for the massive rock excavation in the tunnels and underground powerhouse cavern. This project was built by Procon Mining, the Tahltan Nation Development Corp., along with Formula Contracting and Westpro Contracting, and as far as I know, it turned out to be a very successful project for all involved.

After eventually receiving the contract award for the Long Lake Hydro project in June, the team got into high gear to complete mobilization and work plans so we could get started on site as quickly as possible, as most of the winter's snow was now melting quickly. Art went up to the site to manage the mobilization and get the concrete batch plant set up, but the plan was that he would not stay on the project for very long. WVCC assigned Dan Cave as project manager as well as John Carvel as area superintendent and Rob Wood as project engineer. Both John and Rob were relatively new hires but both had good experience and showed much promise. EBC assigned Jose Rochette to the project as project manager, along with Yves Gagnon as a general superintendent and Patrick Chandonia as a project coordinator, while Neilson assigned Dominique Hotte to oversee the rock excavation and tunnel work in the intake area. Work on site was getting underway in early July and there were some growing pains working matters out between the three new joint venture partners, so much so that Art, yet again, had to come in and assist to get the first season of this project off to the best start possible.

Interestingly enough, on May 1, 2011, while I was in the middle of transitioning the company with Tyam ownership, starting the new working relationship with EBC/Neilson, and getting the Long Lake Hydro Project underway, I separated from my second wife, Gladys, after 13 years of living together, 10 years of that married. A legal separation agreement was drawn up fairly promptly, the house was sold by October and we went in two different directions. We had a lot of great times together, but in the end, Gladys wasn't really happy and that made me unhappy, so, sadly, it was over.

The Long Lake project was originally a three-season project, mainly due to the severe winter weather, extreme heavy snowfall and steep, mountainous terrain, but even though we budgeted to complete the work in three years, it was our plan to get it done in two years, if everything went well. EBC and Donald Proulx, in particular, were big believers that it should not take any more than two seasons to build. In fact, I clearly remember Donald standing up once in a management meeting in our board room with his arms spread wide exclaiming, "This project is easy!!" (To me, being cocky like that in this type of construction is breaking a Cardinal Rule, because things can very easily take a serious turn for the worse.) Between the three companies involved in the joint venture, we had a lot of experience, financial capacity, equipment and horsepower, but we also had some issues that we needed to manage carefully. We made reasonable headway that first season, even with a bit of a bumpy start, but there was clearly no way we would finish the project the next season. One of the biggest problems on the job revolved around Yves Gagnon, the general superintendent nominated by EBC/Neilson. The best way to describe this situation was that Yves may have suffered from a serious case of Napoleon complex (small man complex), and he did everything possible to engender himself as a big-time operator. He had his fingers into every corner of the project, essentially disrupting the work and overriding every decision other foremen or area superintendents made rather than coordinating the work with them. He was also spending money like a drunken sailor on shore leave. He rubbed many of the staff the wrong way, mostly because he thought he was in charge of the entire project, which was not the intent at all. Eventually, after wreaking havoc on the job that first season, Yves was removed from site by October 1, 2011, along with Dominque Haute, who was clearly having trouble adapting to the challenges of this project in Northern B.C., where we relied heavily on his expertise and experience to undertake the rock and tunnel work at the intake structure. We found out some time later that Yves' biggest claim to fame was working on a large hydro dam in Northern Manitoba, but it turned out that was a cost-plus contract, not a hard-dollar contract like the Long Lake Project. (On a cost-plus contract, there is very little measurement of efficiency, and essentially, the more the contractor spends, the more profit the contractor makes!)

Many aspects of the work were well underway after the first season, and we made some good headway sorting out some of the staff positions, specialty subcontractors and equipment required to perform the work efficiently, so we were in a much better position to get the work going in 2012. Sometime that fall, I set up a large, highly technical organizing meeting with all of the subcontractors, major suppliers and key management so we could carefully step through the remaining scope of work in great detail and build an overall project work plan, a reliable completion schedule and a detailed budget to complete the work. I recall that the boardroom was chock full with at least 12 people attending and the meeting went extremely well. Mike Luers sat at that meeting attentively tapping on his laptop throughout the entire day. As I was chairing this meeting and guiding everyone through the various topics, I wasn't able to take comprehensive notes, so after the meeting I asked Mike if I could borrow his notes from the laptop to assist with accuracy of the minutes of the meeting. It was important to document the discussions accurately so everyone was operating on the same wave length following the meeting. Mike responded that he would check them to see if there was anything of value and get back to me. The next day, he reported that there really wasn't anything of use, so I politely asked him why he attended such a long, strictly technical meeting in the first place. He responded that he needed to learn my job so when I eventually retired, he could take over my job. I took Mike aside and as politely as possible, told him quite honestly that he would never learn my job as I had been working in construction since I was 14 years old and he had just started a year ago. I told Mike that he was a very capable executive vice president, but he would need to ensure that experienced construction people ran the nuts and bolts of the company, particularly for complex projects such as these private hydroelectric projects. I was never sure how Mike took that comment, but I meant no disrespect and thought it was valuable advice for him to understand.

One of the strict business fundamentals I have always strongly believed in is that every company requires an organization chart that clearly and accurately represents the various levels of authority and responsibility and establishes correct lines of communication. While this structure seems natural to many, this principle was strongly reinforced by Paul Ridilla,[43] the travelling construction guru, who was our guest speaker at WVCC many years earlier. To this day, I can clearly recall stories Paul told about how this type of chart not only had to be accurately produced and communicated to all employees, it had to be followed, otherwise it was useless. This organization chart was produced for the new WVCC organization as well as the Long Lake Project. A bit later in the construction season, I received an email string from Art at the site that originated from Mike Luers. Art's concern was that he was now receiving orders directly from Mike, the Executive VP of Tyam, and he had instructed Art that he would be travelling to the site later that week and set out a detailed, hour-by-hour schedule for Art to personally tour him around and explain everything

43 During the writing of this memoir, I read a book written by Arnold Palmer called Arnold Palmer – Memories, Stories and Memorabilia from a Life On and Off the Course. This was a nice hardcover book I had originally given Stan Stewardson in 2004. Stan's wife, Heather, called me recently and said she wanted to give this book back to me, since Stan had passed away a few years earlier. On page 148 of that book, Arnold wrote that Paul Ridilla was a good friend of his and that Paul was a member of the Latrobe Country Club! How is that for a wild coincidence?

that was going on. I thought this was quite odd because I knew nothing of any planned site visit. Art reported to me, not Mike and, interestingly enough, I wasn't even copied on this email. I was obviously quite frustrated by this inappropriate intrusion (as Art was) and failure to operate within the intended framework of the company's organization structure. I responded to that email string directly to Mike with a copy to Rob Barker stating roughly that "*we didn't need any Nimrods going up to the site pestering Art, as he was busy enough without this unnecessary intrusion.*" I cannot recall if that site visit actually occurred but the Tyam folks were starting to become quite problematic and this example was only the tip of the iceberg. Work on site ground to a halt in late fall due to the onset of heavy snowfalls and our entire team started concentrating on a much-expanded plan for the 2012 season.

Earlier that season, in October of 2011, I was approached by Barnard Construction from Bozeman, Montana to possibly work together as a joint venture on the Ruskin Dam Rehabilitation project that had recently come out for tender by BC Hydro. This was a good-sized heritage hydroelectric facility that was operated by BC Hydro near Mission, B.C. that required a major upgrade. It just so happened that Barnard Construction was a fairly large, family-owned company that had extensive specific experience in this type of dam retrofit project where the work was required to be underway during the operation of the plant. I recall the value of this project was approximately $150 million. This model fit beautifully into our corporate business plan, and I discussed this opportunity at length with Mike Luers and Rob Barker, who were keen to proceed. Barnard and WVCC eventually formed a joint venture where WVCC was a 25% partner with Barnard carrying 75%. Again, Robin McFarlane and Mike Demers assisted to assemble the joint venture agreement while Lesley Huygen, Scott Ashton and our external accountants sorted out the surety and security requirements with our bonding agent Brian Lawson.

Rainer Kraft, Alex Morrison and I worked on the bid for WVCC, while Jeff Higgins and Paul Kraus worked with several of Barnard's staff to estimate the work from their office. An initial site visit and pre-bid discussions took place in our office in Langley in early October. Once the preliminary bids were nearly complete, along with the proposed schedules and work methodology write-ups, Rainer, Alex and I flew down to Bozeman in early November to combine the bids together and assemble the initial draft of the proposal. I was very impressed by the Barnard organization throughout that process. They had a great blend of some very experienced field people involved throughout the entire estimating and proposal process along with the necessary mix of senior executives and in-house specialists. One of those specialists was a very talented scheduler who worked with us through every step of the estimate, and whenever we made any change to the methodology, production, crew size or order of the work, he was able to make those changes immediately to the schedule and display the effect on one of the big screens in the "war room". All of the WVCC people worked extremely well with Barnard and the bid was a great joint effort. On a couple of occasions where we were discussing critical work, I believe it was the owner, Tim Barnard who sat in on the review and asked some key questions.

Over the next week or so, we were able to complete all of the required changes to the bid and proposal and everything was signed off on the evening before the bid closing date. I had earlier scheduled my third knee surgery, this time for a torn meniscus, on the exact day of the bid closing, but luckily everything was completed and all signed off the day before. All we had to do was drop off the proposal at BC Hydro's head office in Burnaby. I showed up at the Langley hospital that morning and the arthroscopic surgery went ahead as planned. I think I had an epidural block, so I was out of commission for some time after the operation, even though I was awake for the procedure (as best as I can recall). Once I recovered sufficiently and got out of the hospital, I opened up my cell phone and found a flurry of emails between Mike Luers and Barnard Construction, so something went seriously awry! Once I dug deeper into this matter, I learned that Tyam, our parent company, cancelled the JV agreement with Barnard and pulled our involvement in the bid at the last minute. I called Mike Luers in a panic, and he explained to me that both Rob Barker and Ralph Jordan, the two financial partners of Tyam, became uncomfortable with the risks associated with this bid and decided to back out. Mike told me their discomfort arose due to the fact that the contractual risks and liabilities shared by the joint venture parties were "joint and several". Wikipedia defines this as follows:

> "Under **joint and several liability** or all sums, a claimant may pursue an obligation against any one party as if they were jointly liable and it becomes the responsibility of the defendants to sort out their respective proportions of liability and payment."

In essence, this means that if the JV ran into a complete financial disaster on the project, BC Hydro could force the bonding company to complete the project under the terms and conditions of the surety agreement and general contract. The bonding company would then, in turn, take whatever actions necessary to recover the whole amount of the damages it suffered to complete the project from either one or both of the JV parties. They would not try to collect 75% from Barnard and then collect the remaining 25% from WVCC, just because of the makeup of our JV agreement. The most likely scenario, of course, was that the bonding company would go after the party with the deepest pockets, and that was clearly Barnard, not WVCC. In practice, the joint venture agreement stated that WVCC could only be liable for 25% of the losses the JV incurred, so even if the bonding company went after all of the damages from WVCC for some reason, WVCC would have every legal right to collect 75% of those costs from Barnard. However, one potentially possible scenario was if Barnard went bankrupt and the JV was not able to complete the project, WVCC could end up carrying a greater share of those project liabilities. Our protection, regardless of the bonding arrangements, was based on the long-standing strength of Barnard, their strong financial balance sheet and their extensive experience with this type of work. They were a very well-established company that completed around $400 million worth of work annually, so the likelihood of extreme financial hardship or bankruptcy was extremely unlikely, and we were comfortable with that risk. Joint and several liability is common in the industry, and it is a measured, known risk that companies participating in joint ventures need to consider and manage appropriately. It was shocking to me that these businessmen (Rob Barker, Ralph Jordan, Mike Luers and Scott Ashton) were not aware of this common practice when they decided to enter into the heavy construction industry and form this type of joint venture!

Needless to say, I went ballistic! Nobody tried calling me to ask my opinion; they just acted in a needless panic. I told Mike I was completely pissed off with their hasty action because they simply did not understand this was just a qualified proposal to begin negotiations with BC Hydro, not a legal offer to enter into the contract, like many tenders. Even though there were surety arrangements put in place for the joint venture, I don't even think bonds had to accompany the proposal. The bottom line was that we could have easily gotten out of the proposal process after the submission date. I told him we had probably spent $100,000 on putting the proposal together and this had been a great opportunity to work with Barnard in a risk-sharing joint venture arrangement that followed our business plan to a tee. Now they had pulled the rug out from below Barnard and all of our team at the 11th hour! After this final fiasco, I told Mike that once my two-year employment contract was fulfilled that coming August, I would not be staying on with the company as there were too many of these disruptions preventing us from being successful. I had always told Mike and Rob that I was not in a rush to retire, and as long as we were making money and having fun, I would stay on, but this changed all of that. I wanted them to be prepared so it didn't come as a complete surprise to them in August.

Shortly after that call with Mike, I called Tim Barnard, the owner of Barnard in Boseman, and apologized for what happened. I told him I felt terrible about this fiasco and the difficulty it caused them to get their own proposal in without WVCC. I went on to tell Tim that I really appreciated working with everyone at Barnard and had been really looking forward to working together on that project, but there wasn't much I could do at that point. Tim also told me that he appreciated all of the good work that Rainer, Alex and I had done with them on this challenging bid, and he respected everyone from me down, but he was furious with the people above me. I couldn't disagree.

While all of this was going on, our team was working hard with EBC/Neilson to finalize all of the plans for the Long Lake Project for 2012 and a two-day meeting was scheduled for the upcoming weekend. Mike Luers got hold of me and said that the ownership was aware that I was unhappy about the most recent turn of events with the Ruskin Dam proposal and that they were having a meeting of the Board of Directors early that week to discuss that matter, and they would meet with me on the morning of Thursday, January 12. I prepared myself for this meeting, as usual, to be as effective as possible, and I made one sheet of bulleted notes on what I felt management needed to do and not to do to make this company successful. Thursday morning came along and I attended in the boardroom where Mike Luers, Rob Barker and Scott Ashton had seated themselves all down one side of the table. I selected a seat opposite them and the meeting was chaired by Mike Luers. Mike was the first to speak, and he restated that ownership knew I was unhappy and said they had discussed the matter and determined the best way to deal with that was to terminate my employment forthwith.

It felt like I was in a room where the air had been immediately sucked out of it. My lungs felt the void and my heart may have skipped a beat or so. This was shocking news to me! These Nimrods were letting me go from a company Paul and I started and had successfully operated for 25 years?? I waited a full minute, completely expressionless, lightly tapping my fingers on the boardroom table, then I calmly asked two questions.

"You mean I don't need to come to work tomorrow morning?"

Mike quickly answered, "No."

"You mean I don't have to chair the joint-venture meeting for Long Lake this weekend, which I have been working on for the last two weeks?"

Mike answered confidently, "No, we have that covered."

I waited another full minute, digesting this information and still lightly tapping my fingers, all the while thinking that we all knew they had to pay for the next six months at $20,000 per month as a term of the sale agreement and my employment contract, but I didn't have to come to work???

This was starting to sink in and make some sense. Paul and I had received full payment for WVCC and our ongoing payments for Westpark were very secure, one way or another. I stood up suddenly, smiled the biggest smile that I had for a long time and looked each of them in the eyes and said, "Boys, it doesn't get any better than that!" I reached over and shook each of their hands. Their eyes were wide open and their faces were pure white. As I walked out of the room, I could only wonder to myself, *What could they have possibly been thinking?* But if they wanted to pay me $20,000 a month for the next six months to stay away, I could actually live with that concept. It was better than dealing with them for the next six months. I immediately walked downstairs to Art's office, closed the door and told him that they canned me. He was aghast! "Who is going to run the company? None of them has any ability to do that!" I think Art was as shocked as I was and probably wondered how this was going to affect his life. I then walked over to Lesley's office, closed her door, told her what happened and she said exactly the same thing as Art did. I think she went one step further to say, "This isn't going to work." My next question for Lesley was whether she had ever wanted to visit Australia. Her response was that if she was going to fly that far away, she would want to see New Zealand as well.[44] I said I had no problem with that and that made perfect sense to me. Needless to say, Lesley was the first of many to hand in their notice of resignation.

Mike Luers had been stalking me after I left the boardroom to escort me to the front door, as they do in the movies, so I wouldn't steal any trade secrets or anything. Once I exited Lesley's office, he motioned me to my office so I could collect my personal belongings and then stayed with me every minute for the next hour or so as I moved all of my personal files off the computer onto stick files, gathered all of my personal books, etc. into

44 I should inform the reader that at this time I had been dating Lesley, our company accountant, for the last few months.

a cardboard box and loaded it onto a cart. I was out the door in no time. Good thing I didn't have a company car or I would have had to call a taxi! I think they had my final cheque and employment papers prepared by the next day.

That afternoon, I headed to a regularly scheduled meeting in Burnaby with the contractors' association to continue negotiations with BC Hydro on the development of their new construction contract. Scott Jacobs, whom I had known and highly respected for years was on that committee, and I remember him being as shocked as I was when I told him I had been canned earlier that day. Scott had just started Jacob Bros. Construction and it didn't take long for him to suggest we get together and talk about the possibly of working together. This was a great potential working relationship as I really liked Scott, but I have to admit, by this time, I was really liking the idea of taking a well-earned break and getting six months off with full pay and no stress!

Six days later, on January 18, 2012, I purchased two return business class airline tickets to Melbourne, Australia with stopovers in Auckland, New Zealand. Around that same day, I called Bryan Hall to tell him we had something in common—we were both fired by Rob Barker and Mike Luers from the respective construction companies that we started! He howled with laughter! We met that day at Earl's in Langley where we had some nice appies and a few glasses of wine, along with some great stories and more laughs. Shortly after that meeting, I headed down to Palm Springs to play some golf and get used to my semi-retirement. After that famous January 12, 2012 meeting, I had several conversations with various staff employees of WVCC and virtually all of them said there was no future at WVCC and they all left, one by one. I think in retrospect, Mike Luers and Rob Barker thought that they should just let EBC/Neilson build the Long Lake Project and save all of its own overhead costs. After all, those companies had annual volumes of between $500 and $750 million a year combined, so what did they need Rick for?

When the dust settled on the Long Lake Hydro Project, the joint venture lost a total of $55 million on a $76 million job! Donald Proulx told me this when I ran into him at an Independent Hydropower convention in Vancouver some time after the project was completed. WVCC's assets were auctioned off and the company died, just like that.

Just after being terminated from WVCC, and while I was finalizing some extensive travel plans, I received a call from Andrew Purdy, the owner of Ruskin Construction, another general contractor from B.C. He heard that I was no longer working and wanted to know if I would consider working for them on the Deh Cho Bridge, which crossed the Athabasca River in Northern Alberta. I had heard through the industry that they had some serious issues on that bridge with ice flows overtopping a cofferdam and washing out their work bridge, etc., and it sounded like a very interesting and challenging project for a younger man. Andrew attempted to entice me by saying he would pay me $30,000 to $35,000 per month to work there for the next six months to finish off the bridge. I was flattered that he thought so highly of me, but I was almost (unintentionally) rude to him by laughing and saying that was absolutely the last thing on my mind to do. I told him I had just retired from 40 years in construction and I was looking forward to travelling and taking it easy. I wished him luck finding the right guy to finish that job for him.

Just before our departure date to Australia, we had a bit of a celebration at Jimmy Mac's Pub, which was located very close to our old WVCC office in Langley. We called the party the "Hasta La Vista" celebration as my final parting shot to the entire Tyam fiasco. A lot of friends and employees showed up, and we had a great night sharing drinks and appies.

Just after returning from New Zealand and Australia, where Lesley and I travelled for nearly three months in the spring of 2012, I received a call from either Mike Luers or Rob Barker. He told me they wanted to get out from the remaining term of the Westpark earnout agreement. Their concern was that they had zero control over their share of Westpark for another three years, even though they were effectively 50% owners of the company. For obvious reasons, they were concerned we could drain the company of its assets for the next few years and leave them nothing. You might recall I mentioned previously that a key provision in the earnout sale agreement prevented the purchasers from having any voting rights until the shares were fully paid for, and this was now belatedly recognized by the purchasers as a serious obstacle. This was indeed a very interesting situation, particularly after Westpark had two very successful years and Paul and my initial payments for the shares were maxed out based on Westpark's stellar financial performance. But an even more interesting aspect of this situation was the question: Because Westpark had made a considerable profit since the sale agreement was executed, who was the rightful owner of Tyam's share of those profits if they did not conclude the sale over the five-year earnout period? One could argue that if the sale was not completed by Tyam, then perhaps those profits would revert to the seller, namely Paul and Rick. (Imagine me rubbing my hands together like Ebenezer Scrooge!) I discussed this very interesting and potentially lucrative matter with my accountant, Paul Martin, and Robin McFarlane, and we started to develop a plan. Rightly or wrongly, Paul and I really did not wish to repurchase the Westpark shares personally at that time. This had nothing to do with any lack of confidence in Westpark's ability or consistent profitability, rather, both of us had just gone through a lot of work and stress completing the sale of WVCC and our appetite was just not there to get back into a related business that we had just sold. The concept that was eventually envisioned, mainly by Paul Martin, was that the absolute cheapest way to get those shares into someone's pocket was for Westpark to purchase those shares directly with proceeds of the company. The actual evaluation of those shares and the use of pre-tax money by Westpark was literally a dream to Paul Martin, who was ecstatic about this possible deal due to its tax benefits. When we approached D'Arcy and Bill Soutar with this idea, they were both very interested. After they sought independent legal and accounting advice and we spoke to Robin McFarlane further about this plan, we all agreed this was by far the best way to proceed. In due course, D'Arcy and Bill agreed to buy out the earnout agreement from Tyam for a discounted price and continue making the annual payments to Paul and me in accordance with the terms of the original sale agreement, so we were no better or worse with the change. On the other hand, Westpark would be way ahead and now rightfully own 100% of the company shares. For this advantage, and for my brokering of this deal, I was paid a fee from Westpark in the form of a non-competition agreement, which helped me bridge financially into the retirement world. Westpark continued to be very successful through the next few years to completion of the earnout agreement, so Paul and I maxed out our payments and both Westpark and ourselves are now totally protected from any liabilities from the entire Tyam fiasco.[45]

After travelling for most of 2012, I began providing consulting services to a number of contractors and owners on a part-time basis, using my years of experience in construction management. Many of these opportunities often involved disputes or litigation involving construction, and I soon started to build a resumé for acting as an expert witness in many of these litigations. This work continues to be remarkably enjoyable and rewarding because I have historically enjoyed these type of contract issues, but it is so much better when these disputes do not have the possibility of crippling you financially, which was the case when I worked on WVCC disputes. Because I can do most of this consulting work remotely, I relocated to the Okanagan and built a new home overlooking Skaha Lake just south of Penticton where I am able to enjoy the slower pace while still serving many of my clients operating in the Greater Vancouver area.

In addition to consulting over the last 10 years, which involves working about 30–40% of my time, I continue to enjoy a number of activities to keep me busy, including my membership at Fairview Mountain Golf Club in Oliver and my membership at the City Centre Gym in Penticton as well as hanging out with Mary and our many friends and family in the sunny Okanagan. I also love to see Taralyn, Collin, Chaylene, Shannon and my grandchildren Cassidy, Weston and

45 An interesting side note to the Tyam fiasco, Ralph Jordan sued Rob Barker in the Supreme Court of BC on September 20, 2013 over numerous aspects of their business dealings over the last several years including the operation of WVCC since its purchase! We were truly fortunate to be rid of that bunch!

Poppy as much as I can as they are growing up. I have also more time for things like camping in my truck and camper, quadding in the mountains in the Kootenays with Mark Sinclair and Vic Seder, trout fishing in Thalia Lake with Art, Vic Seder and Kerry Grozier (even though we usually only catch one fish per trip!), salmon fishing at Hakai Pass every August with our regular group, boating on Okanagan Lake, going on motorcycle road trips and annual trips down to Palm Springs for winter golf and going on my annual motorcycle trip to Cabo San Lucas.

Salmon Fishing at Hakai Fishing Club

Ten years after the sale of WVCC, my long-time business partner and good friend, Paul Manning, sadly died of Huntington's disease. Huntington's disease is a rare genetic disease that causes the progressive degeneration of nerve cells in the brain and has an extensive impact on a person's movement, coordination and cognition, which usually results in loss of functional abilities over a period of time. Prior to the sale of WVCC, I noticed Paul had some minor facial twitches and body mannerisms, but no one thought much about these at that time. One other thing I noticed over the last several years working with Paul was that he seemed less and less willing or able to deal with stressful situations. This can be very difficult in a dynamic company such as WVCC and this, in part, lead to our eventual decision to sell the company while we were still both in our mid-fifties. In retrospect, we can all see that Paul's symptoms were a result of this disease. With the 20/20 vision of hindsight, I can certainly see the dramatic changes that Paul experienced prior to the sale and how they continued to escalate in the years following. I am glad that Paul had some very good active years after the sale. He certainly did not go down without a battle, and he and his wife Jane were involved in ground-breaking research and fundraising for this terrible disease. I know Paul would have loved to be involved in the writing of this book with me, and he would have really enjoyed the many memories I dredged up from our working together for 25 years. It would have also been great for me to have him involved as he always had a very keen memory for both names and details, far better than mine!

Paul and Me on One of Our Many Fishing Trips at Hakai Land & Sea

One thing I am thankful for is that I was given so many great opportunities as a young man to become involved in a number of larger bridge projects around the province like the Taghum Bridge, Lillooet Bridge,

McPhee Bridge, Carnes Creek, Alexander Creek, then later the Castle River Bridge and Ashcroft Bridge. These were all a nice-sized projects that all had their special challenges but were a lot of fun to be involved in. They certainly gave me a lot of valuable bridge experience to build on. On the flip side, one small regret I have is that I never worked on a real big bridge job like the Alex Fraser or Port Mann Bridges. I was actually offered a job by Hyundai on the Skytrain Bridge between Surrey and New Westminster in the mid-80's but this was just as we were starting WVCC, so I declined that offer. Although those type of jobs would be completely different with large multinational joint-venture teams building them, I think it would have been a good experience for me.

In the end, I believe I had a good career overall, made a decent living and had a lot of fun and great experiences doing it. I started at the bottom in construction as a labourer and worked my way up through my experience as a surveyor, painter, inspector, assistant superintendent, superintendent, general superintendent, project manager, contracts manager, general manager and owner. Once we started WVCC, the company grew quickly, and we accomplished a lot in our early years. WVCC did not grow to its fullest potential by any means, partly due to my own conservative nature, but in my opinion, we always maintained a manageable volume to best service our clients with a type of work to keep our main people busy, and we were always profitable. One larger company that was looking at possibly purchasing WVCC asked us why WVCC wasn't 10 times bigger than it was. Our response was that WVCC was the size that we wanted it to be where we could still enjoy our lives, our families and our hobbies. One thing we were always fortunate with at WVCC was our dedicated, long-term staff and our many hard-working and multi-talented key hourly employees. Everyone pitched in and worked hard to make WVCC successful, and I certainly appreciate the extra effort all of those people gave. It felt like it was a family!

Being peripherally involved in the heavy construction industry over the last 10 years through my consulting business and witnessing first-hand how dramatically construction has changed, my favourite saying has become:

"I'm glad that I built shit in the 70's and 80's!"

THE END

EPILOGUE

THIS STORY I HAVE WRITTEN is based on my memory of the facts of my life without any intentional confabulation or embellishment. Having been terminated from WVCC, I left with virtually no printed or digital records including photographs from those 25 years, so my reference material to rely on was quite limited. However, with the help of many people who I connected with during the writing of this book, the pictures became much clearer and the names began flooding back into my memory bank. That experience was extremely gratifying as well as emotional, and I thoroughly enjoyed the entire process of writing these memoirs. I would highly recommend it to anyone— everyone has a story!

IN MEMORIAM

THIS BOOK WAS WRITTEN IN memory of Ersul Fox, Clem Buettner, Paul Manning, Bill Tough, Stan Stewardson, Ed Terris, Dan Schweers, Dave Manning, Ken Sleightholme, Don Evers, George Kinakin, Wayne Strobbe, Doug Mace, Wayne Clark, Dennis Case, Glen Mosely, Karl Schlauwitz, Fred Dickhut, John Jones, Chris Nichols, Ed McKenzie, Art Forster, Jim Goodbrand, Pat Charette, Bernie Lofstrand, Joe Depedrina, Lloyd McHarg, Larry McKee, Ron Elliott, Barry Dong, Gus Angus, Bob Bruce and Arnold Talbot.

There are many great people who are sadly missed in this list!

ACKNOWLEDGEMENTS

THE FOLLOWING PEOPLE ASSISTED ME to correctly remember and improve the accuracy of the book. This was a great chance to touch base with so many people from my past, and I am indebted to all these friends and acquaintances with better memories than mine:

Shawyn Corbett, Tom Beck, Hugh Brown, Mike Senkiw, Jeff Mullins, Kerry Grozier, Dan Neil, Vic Sedar, Rick Hanson, Kathy Tough, Jim McKenzie, Mike McNeill, Kathleen Oppio, Shirley Beasom, Patty Palesch, Bonny Jaggard, Doug MacLanders, Don Clipperton, Steve McAlister, Roy Buettner, John Simonett, Kris Thorleifson, Bob Burns, Art Penner, Drew Copley, Rocky Vanlerberg, Stu Graham, Vern Dancy, KC Shenton, Pete Steiner, Dan Cave, Joe Mather, Al Beamer, Barry Dong, Ray Schachter, Marny Woolf, Paul Schlauwitz, Gary Gallagher, Jon Lee Kootnekoff, Vera Kinakin, Sara Kinakin, Dave Coutu, Bruce Lowe, Al Northrup, Blair Squire, Vic Bilodeau, Ron Kruhlak, Q.C., Philip Hochstein, Bob Heath, Tom Bayntum, Scott Bradley, Dale Lauren, Mike Demers, Lewis Clarke, Bill Soutar, Paul Kraus, Bryan Hall, John Hope, Jim McGill and Scott Jacobs.

I wish to provide my sincere thanks to Brian Taylor, who has been a good friend for many years who generously offered to read the book and provide his constructive comments once the first draft was completed. I owe particular thanks to my sister Kathy for all of her determined investigative work on our family ancestry, which I was able to use in the book. Next, to my fiancé, Dr. Mary Kjorven, who I am deeply indebted for the many, many hours she spent in her already busy days to provide her excellent and tireless editing skills and objective feedback. Essentially, I wrote the book from my mind and memory, perhaps the way one might verbally tell a story, but Mary adapted this manuscript to proper English and diction. Finally, I owe a debt of gratitude to Kelly Jardine of Clearly Read Editing for the final editing of the manuscript. It was a pleasure working with Kelly and I would highly recommend her services.

Taralyn's Wedding Day – August 2, 2009

Chaylene's Wedding Day – October 26, 2016

MY PROUDEST ACHIEVEMENTS BY FAR

CPSIA information can be obtained
at www.ICGtesting.com
Printed in the USA
BVHW060724230722
642745BV00003B/6